工业和信息化精品系列教材
网络技术

Network Technique

华为版

网络互联
技术与实训
任务式微课版

甘卫民 周化 ◉ 主编
李毓丽 张明军 毛锦庚 刘思怡 ◉ 副主编

人民邮电出版社
北京

图书在版编目（CIP）数据

网络互联技术与实训：任务式微课版：华为版 / 甘卫民，周化主编. -- 北京：人民邮电出版社，2024.7
工业和信息化精品系列教材. 网络技术
ISBN 978-7-115-64426-8

Ⅰ. ①网… Ⅱ. ①甘… ②周… Ⅲ. ①计算机网络－高等学校－教材 Ⅳ. ①TP393

中国国家版本馆CIP数据核字(2024)第095973号

内 容 提 要

本书以网络互联技术为主线，采用"学习目标—知识准备—任务实施—任务拓展"的结构体系进行讲解。全书共 6 个模块，包括网络设备和仿真平台、交换机组网技术、路由器互联技术、广域网技术、网络安全技术及网络服务技术。全书设置 13 个任务，所有任务的内容均基于华为网络仿真工具平台 eNSP，使用华为命令实现。

本书可作为高等院校计算机网络技术专业及其相关专业对应课程的教材，也可以作为相关培训人员或工程技术人员学习路由交换/网络互联技术的参考书。

◆ 主　　编　甘卫民　周　化
　副 主 编　李毓丽　张明军　毛锦庚　刘思怡
　责任编辑　顾梦宇
　责任印制　王　郁　焦志炜

◆ 人民邮电出版社出版发行　北京市丰台区成寿寺路 11 号
　邮编　100164　电子邮件　315@ptpress.com.cn
　网址　https://www.ptpress.com.cn
　固安县铭成印刷有限公司印刷

◆ 开本：787×1092　1/16
　印张：15.75　　　　　　　　2024 年 7 月第 1 版
　字数：413 千字　　　　　　　2025 年 3 月河北第 2 次印刷

定价：59.80 元

读者服务热线：(010)81055256　印装质量热线：(010)81055316
反盗版热线：(010)81055315

前言 FOREWORD

通信技术日新月异，全新 IP（互联网协议）技术已成为下一代网络技术发展的总趋势。计算机网络技术作为信息技术的核心，由于其实现技术和方法的不断更新，推动着信息技术产业的螺旋式发展。在这种背景下，服务器中的数据如何高效、高质地为网络边缘的用户提供更好的支持？从服务器到用户终端的数据包传递与转发如何经过不同的网络设备和链路？这就需要大家都遵守一系列转发规则，即网络协议，包括子网内的交换协议和子网间的路由协议。学习完网络互联技术后，读者会找到以上问题的答案。因此，网络互联技术是计算机网络技术及其相关研究方向的核心和基础，读者掌握这门技术对于后续课程的学习具有重要意义。网络互联技术可作为计算机网络技术专业核心课程的前序课程，在网络人才培养中起到重要作用。

本书利用华为 eNSP 模拟真实网络设备运行场景，使读者能快速熟悉华为数据通信系列产品，了解并掌握相关产品的操作与配置方法，提升对网络的规划、建设、运维能力，从而构建更高效、更优质的企业网络。本书围绕网络互联工程的实施，设计了一系列的工程任务，力求使读者在完成任务的过程中，不仅能掌握职业所需的核心知识和实践技能，而且能获得实际工作经验。

编者利用多年的计算机网络技术的实践教学经验，归纳并总结出网络互联技术的典型案例，围绕数据通信组网综合实训任务，设计各个任务相关技术理论。本书共 6 个模块 13 个任务，通过任务驱动的方式贯穿网络互联技术的基础知识和基本原理，以及操作练习和技能实践。

本书的主要知识点总结如下。

（1）认识 eNSP 界面及相关软件的作用，熟悉 VRP 体系结构，理解 VRP 命令行视图和 VRP 命令级别，掌握 VRP 常用的基础命令，熟悉本地登录管理与远程登录管理的配置。

（2）理解冲突域与广播域相关概念、VLAN 划分基础知识、STP 的原理及链路聚合的原理，掌握 VLAN 划分方法，熟悉 STP 配置及参数的修改配置，熟悉链路聚合的两种模式（手动负载均衡模式与静态 LACP 模式）。

（3）理解路由概念及网络层设备，熟悉 IP 地址、路由表等相关概念，重点掌握路由器的原理与作用，理解路由算法、动态路由协议 RIP 和 OSPF 的工作原理，熟悉静态路由、动态路由、单臂路由、三层交换机路由的配置实现过程。

（4）理解广域网的概念及其封装协议，重点掌握 HDLC 协议与 PPP 的原理与配置方法，熟悉 HDLC 帧与 PPP 帧的结构，掌握 HDLC 协议与 PPP 的典型配置方法。

（5）理解 ACL 的作用、应用场景和 ACL 在企业网络中的应用，了解包过滤工作流程、ACL 匹配的规则，熟悉基本 ACL 命令与高级 ACL 命令的配置过程及在端口下的应用方法，理解 IPSec 封装模式，掌握 IPSec 安全策略在 VPN 隧道中的设计过程。

（6）熟悉私有 IP 地址与公有 IP 地址的应用，理解 NAT 技术的工作流程及 NAT 技术分类，熟练掌握 NAT 技术的几种配置方法（静态配置、动态配置、NAPT、Easy IP），了解 DHCP 的应用场景，掌握 DHCP 全局地址池的配置方法和端口地址池的配置方法，以及交换机端口启用 DHCP 发现功能和 IP 地址分配功能的方法。

本书参考学时为 36 学时~54 学时，建议采用理论与实践一体化教学模式，各模块的参考学时见下面的学时分配表。

模块	课程内容	学时
模块 1	网络设备和仿真平台	2~4
模块 2	交换机组网技术	8~12
模块 3	路由器互联技术	8~12
模块 4	广域网技术	6~8
模块 5	网络安全技术	6~8
模块 6	网络服务技术	6~10
学时总计		36~54

本书由广州软件学院的甘卫民、周化任主编，广州软件学院的李毓丽、张明军、广州南方学院的毛锦庚和四川现代职业学院的刘思怡任副主编，甘卫民编写了模块 1~模块 3，张明军编写了模块 4，周化编写了模块 5，李毓丽、毛锦庚和刘思怡共同编写了模块 6，全书的统稿工作由甘卫民完成。

由于编者水平和经验有限，书中难免存在欠妥之处，恳请读者批评指正。

编者

2024 年 2 月

目录 CONTENTS

模块 1

网络设备和仿真平台 …………1
【学习目标】………………………1
任务 1.1　认识常见的网络设备 ………2
【知识准备】………………………2
 1.1.1　网络互联概述 ………………2
 1.1.2　网络互联模型 ………………2
 1.1.3　网络设备介绍 ………………4
【任务实施】………………………10
 实验 1　双绞线的制作 ……………10
【任务拓展】………………………13
任务 1.2　使用 eNSP 搭建一个简单的端到端网络 ………………13
【知识准备】………………………13
 1.2.1　eNSP 与相关网络软件概述 …13
 1.2.2　eNSP 的安装和启动 …………14
 1.2.3　VRP 概述 ……………………21
 1.2.4　网络设备系统登录管理 ………25
【任务实施】………………………26
 实验 2　搭建一个简单的端到端网络 ……26
 实验 3　配置本地登录管理与远程登录管理 ………………………29
【任务拓展】………………………34

模块 2

交换机组网技术 ………………35
【学习目标】………………………35
任务 2.1　配置 STP ………………36
【知识准备】………………………36
 2.1.1　STP ……………………………36
 2.1.2　STP 的基本参数 ……………38
 2.1.3　STP 的实施过程 ……………39
 2.1.4　STP 的端口状态 ……………41
 2.1.5　STP 的报文 …………………42
 2.1.6　STP 的计时器 ………………44
【任务实施】………………………46
 实验 1　配置 STP …………………46
【任务拓展】………………………54
任务 2.2　VLAN 划分 ………………55
【知识准备】………………………55
 2.2.1　VLAN 相关介绍 ………………55
 2.2.2　链路类型和端口类型 …………57
 2.2.3　划分 VLAN 的基本方法 ………58
 2.2.4　VLAN 注册协议概述 …………59
【任务实施】………………………59
 实验 2　配置 VLAN …………………59
【任务拓展】………………………69

任务 2.3　配置链路聚合技术 ………… 70
【知识准备】……………………………… 70
　　2.3.1　链路聚合技术的概念和作用 ……… 70
　　2.3.2　链路聚合技术的应用场景 ………… 72
　　2.3.3　链路聚合技术的模式 ……………… 73
　　2.3.4　链路聚合技术需要注意的规则 …… 73
【任务实施】……………………………… 74
　　实验 3　配置链路聚合技术 …………… 74
【任务拓展】……………………………… 82

模块 3

路由器互联技术 …………… 83

【学习目标】……………………………… 83
任务 3.1　静态路由的配置 …………… 84
【知识准备】……………………………… 84
　　3.1.1　虚电路与数据报 …………………… 84
　　3.1.2　IP ……………………………………… 85
　　3.1.3　路由的概念 ………………………… 88
　　3.1.4　静态路由的概念 …………………… 91
【任务实施】……………………………… 92
　　实验 1　配置静态路由 ………………… 92
【任务拓展】…………………………… 103
任务 3.2　动态路由的配置 ………… 104
【知识准备】…………………………… 104
　　3.2.1　动态路由协议 …………………… 104
　　3.2.2　RIP 概述 ………………………… 107
　　3.2.3　RIP 原理 ………………………… 110

　　3.2.4　RIP 的防环机制 ………………… 111
　　3.2.5　RIP 的配置命令 ………………… 114
　　3.2.6　OSPF 协议概述 ………………… 115
　　3.2.7　OSPF 报文类型 ………………… 117
　　3.2.8　OSPF 协议的链路状态公告 …… 121
　　3.2.9　邻居和邻接概念的区别 ………… 122
　　3.2.10　OSPF 工作过程 ………………… 123
　　3.2.11　OSPF 基本命令 ………………… 125
【任务实施】…………………………… 128
　　实验 2　使用 RIP 配置动态路由 ……… 128
　　实验 3　使用 OSPF 协议配置动态
　　　　　　路由 ………………………… 142
　　实验 4　配置 VLAN 间的路由 ……… 160
【任务拓展】…………………………… 166

模块 4

广域网技术 ………………… 168

【学习目标】…………………………… 168
任务 4.1　HDLC 协议的配置 ……… 169
【知识准备】…………………………… 169
　　4.1.1　广域网概述 ……………………… 169
　　4.1.2　串行链路的数据传输方式 ……… 169
　　4.1.3　HDLC 协议 ……………………… 170
【任务实施】…………………………… 172
　　实验 1　配置 HDLC 协议 …………… 172
【任务拓展】…………………………… 175
任务 4.2　PPP 的配置 ……………… 176

【知识准备】································176

 4.2.1 PPP 概述····························176

 4.2.2 PPP 的帧格式······················176

 4.2.3 PPP 的工作流程··················177

 4.2.4 LCP 概述····························178

 4.2.5 IPCP 概述··························180

 4.2.6 PPP 的基础配置命令··········181

【任务实施】································183

 实验 2 配置 PPP························183

【任务拓展】································188

模块 5

网络安全技术 ················190

【学习目标】································190

任务 5.1 ACL 的配置····················191

【知识准备】································191

 5.1.1 ACL 概述····························191

 5.1.2 ACL 的应用场景··················192

 5.1.3 ACL 的包过滤流程··············192

 5.1.4 ACL 的类型························194

 5.1.5 ACL 步长····························196

 5.1.6 通配符掩码························196

 5.1.7 ACL 的配置命令和匹配顺序·····197

【任务实施】································198

 实验 1 配置 ACL························198

【任务拓展】································202

任务 5.2 利用 IPSec 的 VPN 技术

     进行配置····················203

【知识准备】································203

 5.2.1 VPN 和 IPSec VPN 概述········203

 5.2.2 IPSec VPN 的应用场景········204

 5.2.3 IPSec 架构··························204

 5.2.4 IPSec 的封装模式················205

 5.2.5 IPSec VPN 的配置流程········206

【任务实施】································207

 实验 2 配置 IPSec VPN 技术········207

【任务拓展】································210

模块 6

网络服务技术 ················212

【学习目标】································212

任务 6.1 NAT 服务器的配置··········213

【知识准备】································213

 6.1.1 NAT 概述····························213

 6.1.2 私有 IP 地址和公有 IP 地址·····213

 6.1.3 NAT 工作流程····················214

 6.1.4 NAT 技术分类····················214

 6.1.5 NAT 基本配置命令············216

【任务实施】································220

 实验 1 配置 NAT Server···············220

【任务拓展】································232

任务 6.2 DHCP 的配置····················232

【知识准备】……………………………232

6.2.1 DHCP 概述 ………………… 233

6.2.2 DHCP 的报文类型和工作

过程 …………………………… 233

6.2.3 DHCP 基本配置命令 ………… 235

6.2.4 DHCP Relay …………………… 237

【任务实施】……………………………238

实验 2 配置 DHCP ………………… 238

实验 3 配置 DHCP Relay ………… 241

【任务拓展】……………………………244

模块 1
网络设备和仿真平台

01

当今时代是一个信息化的时代，无论是在生活、工作还是学习中，网络已经成为不可缺少的部分。网络系统是通过多种网络设备共同构建的，这些设备各司其职、相互配合，组成信息传递的通道。无论是家庭、校园、公司还是其他公共场所，都需要构建一个合适、安全并且完善的网络系统。本模块主要带领读者认识构建各类网络系统所需的网络设备和仿真平台（以 eNSP 为例）。

【学习目标】

知识目标
（1）熟悉互联网络 OSI 参考模型及 TCP/IP 模型。
（2）了解互联网络常用设备，重点了解交换机与路由器的功能及划分。
（3）重点熟悉双绞线的接线及使用方法。
（4）熟悉路由器与交换机设备的 Console 线路与 VTY 线路管理。
（5）了解 eNSP、Wireshark、WinPcap、VirtualBox 软件的作用。
（6）理解 VRP 体系结构。
（7）理解 VRP 命令行的几种主要视图及基础配置。

技能目标
（1）学会常用双绞线的制作方法。
（2）熟悉 eNSP 的安装过程及操作界面。
（3）能够利用 eNSP 绘制简单的网络拓扑。
（4）能够利用命令行进行模式间的切换及基本的参数配置。
（5）能够利用 Console 线路与 VTY 线路进行路由器与交换机的登录管理。
（6）能够理解 VRP 命令行级别的分配权限。

素质目标
（1）增强学生的专业自信，鼓励学生勇于挑战。
（2）培养学生精益求精、勇于创新的工匠精神。

任务 1.1 认识常见的网络设备

【知识准备】

1.1.1 网络互联概述

所谓网络互联,一般是指将不同的网络(如局域网、广域网)通过某种方式连接起来,使之能够相互通信的一种技术,也可理解为一个通过中间网络设备连接多个独立网络的集合,其功能是形成一个覆盖范围更广的网络。

网络互联有 3 个重要概念。

(1)互联:在两个物理网络之间至少有一条物理链路。但这并不能保证这两个网络一定能进行数据交换,因为还要取决于这两个网络的通信协议是否兼容。

(2)互通:两个网络之间可以进行数据交换。但这仅涉及通信的两个网络之间的端到端连接和数据交换。

(3)互操作:两个网络中不同计算机系统之间具有透明地访问对方资源的能力,不依赖于具体连接形式。

1.1.2 网络互联模型

1. OSI 参考模型

国际标准化组织(International Organization for Standardization,ISO)在 1977 年成立了专门机构来研究网络体系结构和网络协议的国际标准化问题。ISO 提出了开放系统互连(Open System Interconnection,OSI)参考模型。这一标准定义了网络互联的 7 层框架,实现了系统间的互联性、互操作性和可移植性。

在 OSI 参考模型中,包括体系结构、服务定义和协议规范 3 级抽象。

OSI 体系结构定义了 7 层框架,从下到上分别是物理层、数据链路层、网络层、传输层、会话层、表示层和应用层,用以进行进程间的通信,并作为一个概念性框架来协调各层标准的制定,如图 1-1 所示。

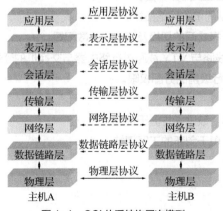

图 1-1 OSI 体系结构层次模型

OSI 服务定义了每一层提供的服务，某一层的服务是指该层及其以下各层提供给上一层的服务，层间的服务通过 OSI 参考模型定义好的层间抽象接口完成，交互时使用服务原语，各种服务不考虑服务的具体实现。

OSI 协议规范说明了控制信息的内容。

OSI 参考模型中各层的特性如下。

（1）物理层

物理层用于建立、维护、断开物理连接，其主要设备是网线、中继器、调制解调器等。

（2）数据链路层

数据链路层的主要功能就是合成数据块，封装成帧与其他计算机进行帧信息传递，实现帧同步、控制差错、控制流量、控制链路、MAC 寻址、区分数据和控制信息、透明传输等服务。数据链路层的主要设备是交换机。

（3）网络层

网络层是 OSI 参考模型中的第三层，介于传输层和数据链路层之间，它在数据链路层提供的两个相邻端点之间的数据帧的传送功能上，进一步管理网络中的数据通信，将数据设法从源端经过若干个中间节点传送到目的端，从而向传输层提供基本的端到端的数据传送服务，在网络层进行逻辑地址寻址，实现不同网络的选择。网络层的主要设备是路由器。

（4）传输层

传输层位于网络层与会话层之间，传输层在终端用户之间提供透明的数据传输，向上层提供可靠的数据传输服务。传输层在给定的链路上通过流量控制、分段/重组和差错控制来保证数据传输的可靠性。传输层有传输控制协议（Transmission Control Protocol，TCP）和用户数据报协议（User Datagram Protocol，UDP），前者用于保证传输的稳定、安全，后者用于保证传输的速度。传输层的主要设备是防火墙。

（5）会话层

会话层位于传输层与表示层之间，其主要功能是会话管理、数据流同步等，如 QQ、微信等。

（6）表示层

表示层位于会话层与应用层之间，向上对应用层服务，向下接受来自会话层的服务。表示层为在应用程序之间传送的信息提供表示方法的服务，它只关心信息发出的语法和语义。

（7）应用层

应用层的功能为"处理"，即通过云计算平台进行信息处理。应用层为用于通信的应用程序和用于消息传输的底层网络提供接口，并提供常见的网络应用服务。

2. TCP/IP 模型

TCP 是一种面向连接的、可靠的、基于字节流的传输层通信协议。互联网协议（Internet Protocol，IP）则是一种面向数据的，用于报文交换网络的协议。互联网的主要协议为 TCP/IP。

因为 OSI 参考模型只是存在于概念和理论上的一种模型，它的缺点是分层太多，增加了网络工作的复杂性，所以没有大规模应用。TCP/IP 协议族按层次划分了五层体系结构或者四层体系结构，大大简化了 OSI 参考模型的某些层次。五层体系结构只是为介绍网络原理而设计的，实际应用更多的还是 TCP/IP 四层体系结构，从下到上分别是网络接口层、网络层、传输层和应用层，如图 1-2 所示。

图1-2 TCP/IP体系结构层次模型

TCP/IP 模型与 OSI 参考模型有清晰的对应关系，覆盖了 OSI 参考模型的所有层次，网络接口层对应 OSI 参考模型的物理层和数据链路层，网络层对应 OSI 参考模型的网络层，传输层对应 OSI 参考模型的传输层，应用层对应 OSI 参考模型的会话层、表示层、应用层，如图1-3 所示。

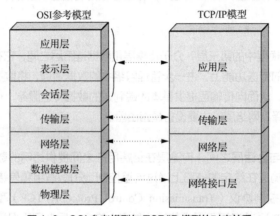

图1-3 OSI 参考模型与 TCP/IP 模型的对应关系

1.1.3 网络设备介绍

1. 计算机设备

（1）服务器

服务器是网络的核心设备，负责网络资源管理和用户服务。服务器如图1-4 所示。

图1-4 服务器

服务器是指在网络环境下运行相应的应用软件,为网络用户提供共享信息资源和各种服务的一种高性能计算机。服务器与普通计算机有很多相似之处,如有中央处理器(Central Processing Unit,CPU)、内存、硬盘、各种总线等,具备高速的运算能力、长时间的可靠性、强大的外部数据吞吐能力等特点,主要用于计算分析、数据存储等。

(2)工作站

工作站是一种以普通计算机和分布式网络计算为基础,主要面向专业应用领域,具备强大的数据运算与图形、图像处理能力,为满足工程设计、动画制作、科学研究、软件开发、金融管理、信息服务、模拟仿真等专业领域需求而设计开发的高性能计算机。

(3)共享设备

共享设备是指为众多用户提供共享的打印机等公用的设备。

2. 网络传输介质

(1)有线传输介质

① 同轴电缆:类型有 RG58、RG11、RG59 这 3 种,传输距离分别为 185m、500m、10km,传输速率分别为 10Mbit/s、100Mbit/s、100～150Mbit/s。同轴电缆如图 1-5 所示。

图 1-5 同轴电缆

② 双绞线:双绞线有屏蔽双绞线和非屏蔽双绞线两种类型。屏蔽双绞线有 3/5 类,传输距离为 100m,传输速率分别为 16Mbit/s、100Mbit/s。非屏蔽双绞线有 3/4/5/6 类,传输距离为 100m,传输速率有 16Mbit/s、20Mbit/s、100Mbit/s、155Mbit/s、200Mbit/s、1000Mbit/s。双绞线如图 1-6 所示。

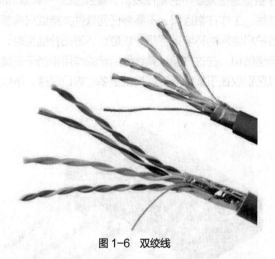

图 1-6 双绞线

③ 光纤:光纤有多模光纤和单模光纤两种类型,传输距离分别为 2km、2～10km,传输速率分别为 100～1000Mbit/s、1～10Gbit/s。光纤如图 1-7 所示。

图 1-7 光纤

（2）无线传输介质

无线传输介质有无线电波、红外线、微波和激光等。在局域网中，通常只使用无线电波和红外线作为传输介质。无线传输介质通常用于广域网的广域链路的连接。

3. 网络互联设备

在物理层，网络互联设备主要有中继器、集线器两种。

（1）中继器

中继器（Repeater）工作于 OSI 参考模型的物理层，是局域网上所有节点的中心，它的作用是放大信号、补偿信号衰减，它支持远距离的通信，起到扩展传输距离的作用，且对高层协议是透明的。中继器如图 1-8 所示。

图 1-8 中继器

（2）集线器

集线器（Hub），属于数据通信系统中的基础设备。集线器是一种以星形拓扑结构将通信线路集中在一起的设备，相当于总线，工作在物理层，不需要任何软件支持或只需要很少管理软件支持。它内部采用了电气互联，当维护局域网的环境是逻辑总线型或环形拓扑结构时，完全可以用集线器建立一个物理上的星形或树形网络结构。在这方面，集线器所起的作用相当于多端口的中继器。集线器实际上就是中继器的一种，其区别仅在于集线器能够提供更多的端口服务，所以集线器又称多口中继器。集线器如图 1-9 所示。

图 1-9 集线器

在数据链路层工作的主要网络互联设备有网卡、网桥和交换机。

（3）网卡

图 1-10 所示为网卡的实例展示。网卡和局域网之间的通信是通过电缆或双绞线以串行传输方式进行的，网卡将计算机的数据封装成帧，并且通过电缆或双绞线等传输介质将数据发送到网络上；而网卡和计算机之间的通信则是通过计算机主板上的输入/输出（Input/Output，I/O）总线以并行传输方式进行的，网卡接收其他网络发送来的帧，并将帧重新组合成数据，通过主板上的 I/O 总线传给计算机。因此，网卡的一个重要功能就是进行串行/并行转换。

图 1-10　网卡的实例展示

世界上每一张网卡上都有一个唯一的编码，称为物理地址（即 MAC 地址），MAC 地址是固化在网卡上的，一般不可改变，是网络上用来表示一个网络硬件设备的标识符。

（4）网桥

网桥是早期的两端口二层网络设备，是用来连接不同网段的设备，同时可以隔离冲突域，因为它的两个端口不是共享一条总线的（分别有一条独立的交换信道），比集线器性能更好（集线器各端口共享同一条总线）。后来网桥被具有更多端口且能隔离冲突域的交换机所取代。网桥只有两个端口，两端所连接的通常是两个物理网段，网桥端口一般不直接连接主机。图 1-11 所示为多端口网桥示意（H1、H2、H3……表示主机）。

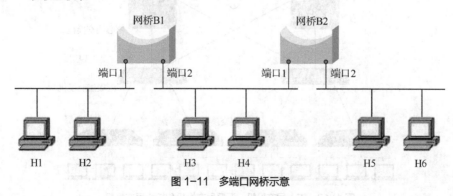

图 1-11　多端口网桥示意

（5）交换机

图 1-12 所示为二层交换机的实例展示。交换机是集线器和网桥的升级换代产品，它既具有与集线器一样的集中连接功能，又具有与网桥一样的数据交换功能，所以交换机又被称为多端口的网桥，其实也可以称作带有交换功能的集线器。它有效地隔离了以太网的冲突域，极大地提升了以太网的性能。

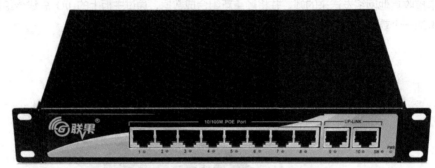

图 1-12　二层交换机的实例展示

交换机的主要性能参数有端口数量、端口带宽、交换容量、包转发率等。

① 端口数量：交换机最直观的衡量因素，通常此参数是针对固定端口交换机而言的，常见的标准的固定端口交换机的端口数量有 8、12、16、24、48 等。另外，部分交换机还会提供专用的上行端口。

② 端口带宽：指交换机端口的数据交换速率，也称为端口传输速率。目前常见的端口带宽有 10Mbit/s、100Mbit/s、1000Mbit/s 等。

③ 交换容量：指整机交换容量，表示交换机内部总线的传输容量。一台交换机所有端口都在工作时，它们的双向数据传输速率之和称为这台交换机的接口交换容量。在设计交换机时，交换机的整机交换容量总是大于交换机的接口交换容量。

④ 包转发率：指一台交换机每秒可以转发数据包的数量，即整机包转发率。一台交换机所有端口都在工作时，它们每秒可以转发的数据包数量之和称为这台交换机的接口包转发率。

交换机的类型按网络构成方式可划分为接入层交换机、汇聚层交换机和核心层交换机，如图 1-13 所示。

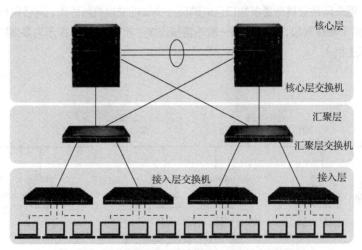

图 1-13　接入层交换机、汇聚层交换机和核心层交换机

在网络层工作的网络互联设备主要有路由器。

（6）路由器

路由器实质上是一种将网络进行互联的专用计算机。路由器在 TCP/IP 模型中又称为 IP 网关，是具有路由和转发功能的实现网络互联的设备，它还有子网间速率适配、隔离子网、指定访问规则的作用。

路由器的主要性能参数如下。

① 吞吐量：指在不丢包的情况下单位时间内通过的数据包数量，也就是设备整机数据包转发的能力，这是设备性能的重要指标。网络中的数据是由一个个数据包组成的，对每个数据包的处理都要耗费资源。路由器吞吐量表示的是路由器每秒能处理的数据量，是路由器性能的直观反映。

② 支持的网络协议：在网络上的各台计算机之间也有一种语言，即网络协议，不同的计算机之间必须共同遵守相同的网络协议才能进行通信。常见的网络协议有 TCP/IP、IPX/SPX、NetBEUI 等。在局域网中用得比较多的是 IPX/SPX。用户如果想访问 Internet，就必须在网络协议中添加 TCP/IP。

③ 线速转发能力：指在达到端口最大速率的时候，路由器传输的数据包没有丢失。线速转发能力是路由器性能的一个重要指标。简单地说就是进来多大的流量，就出去多大的流量，不会因为设备处理能力的问题而造成吞吐量下降。

④ 带机数量：指路由器能负载的计算机数量。路由器的带机数量直接受实际使用环境的网络繁忙程度的影响，不同的网络环境带机数量相差很大。

路由器根据使用级别可划分为接入路由器、企业级路由器和骨干级路由器。

接入路由器主要用于连接家庭或因特网服务提供方（Internet Service Provider, ISP）内的小型企业客户，如图 1-14（a）所示；企业级路由器主要用于连接企业局域网与广域网，如图 1-14（b）所示；骨干级路由器主要用于实现企业级网络的互联，如图 1-14（c）所示。

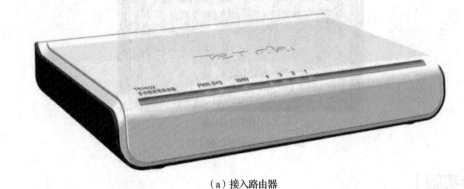

（a）接入路由器

 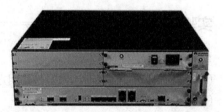

（b）企业级路由器

图 1-14 各种路由器

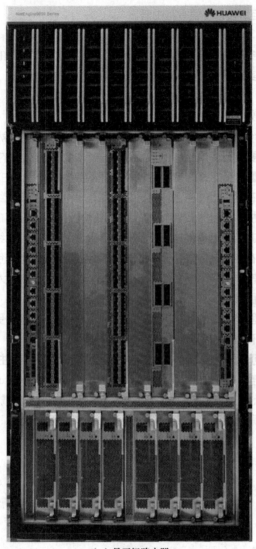

(c)骨干级路由器

图1-14 各种路由器(续)

【任务实施】

实验1 双绞线的制作

实验目的:
(1)了解常见的网络传输介质及其使用方法。
(2)掌握双绞线连接的方法。
(3)掌握双绞线的制作和测试过程。

1. 双绞线的制作方法

双绞线的制作方法有两种国际标准,分别为 EIA/TIA568A(简称 T568A)及

V1-1 双绞线的制作

EIA/TIA568B（简称 T568B）。而双绞线的连接方法也主要有两种，分别为直连法和交叉法。简单地说，直连法就是水晶头两端同时采用 T568B 标准的接法；而交叉法则是水晶头一端采用 T568A 标准的接法，另一端采用 T568B 标准的接法，如图 1-15 所示。双绞线标准如下。

① T568A 标准：白绿、绿、白橙、蓝、白蓝、橙、白棕、棕。

② T568B 标准：白橙、橙、白绿、蓝、白蓝、绿、白棕、棕。

图 1-15　T568A 标准和 T568B 标准

（1）直连法

直连法即双 T568B 接法（常用）：两端的接法一样，线序不变。排列顺序：1 接白橙，2 接橙，3 接白绿，4 接蓝，5 接白蓝，6 接绿，7 接白棕，8 接棕。直连法示意如图 1-16 所示。

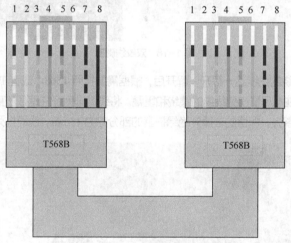

图 1-16　直连法示意

（2）交叉法

交叉法：两端的接法不一样，一端采用 T568B 标准的接法，另一端采用 T568A 标准的接法。交叉法示意如图 1-17 所示。

A 端：1 接白绿，2 接绿，3 接白橙，4 接蓝，5 接白蓝，6 接橙，7 接白棕，8 接棕。

B 端：1 接白橙，2 接橙，3 接白绿，4 接蓝，5 接白蓝，6 接绿，7 接白棕，8 接棕。

2. 双绞线的制作过程

（1）利用压线钳的剪线刀口剪裁出计划需要使用的双绞线长度，剥除双绞线外绝缘护套。可以利用压线钳的剪线刀口将线头剪齐，再将线头放入剥线专用的刀口，稍微用力握紧压线钳慢慢旋转，使刀口划开双绞线的护套，如图 1-18 所示。

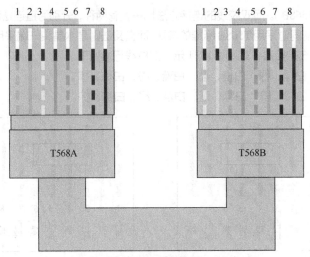

图 1-17 交叉法示意

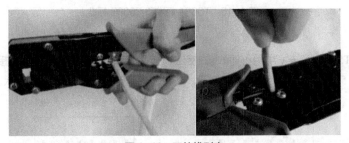

图 1-18 双绞线剥皮

（2）将每对相互缠绕的线缆逐一解开，解开后，根据需要按照 T568A 或者 T568B 标准将线缆排列好并理顺压直，排列时应注意尽量避免线缆的缠绕和重叠。将线缆排列好并理顺压直之后，利用压线钳的剪线刀口把线缆顶部裁剪整齐，保留的去掉外绝缘护套的部分约为 15mm。线缆排列顺序如图 1-19 所示。

图 1-19 线缆排列顺序

（3）将整理好的线缆插入水晶头内。需要注意的是，要将水晶头有塑料弹簧片的一面向下，有针脚的一面向上，使有针脚的一端指向远离自己的方向，有方形孔的一端对着自己。插入的时候需要注意缓缓地用力将 8 条线缆同时沿水晶头内的 8 个线槽插入，一直插到线槽的顶端。使用压线钳压接水晶头，受力之后听到轻微的"啪"的一声即可，如图 1-20 所示。按照相同的方法压接另一端的水晶头。

（4）线缆测试。将线缆两端水晶头分别插入测试仪对应的插口中，观察测试仪指示灯的闪烁顺序，如图 1-21 所示。若任意一芯没有压接到位，则对应的指示灯不亮。若任意一芯线序错误，则对应的指示灯将显示错误的线序。

模块 1
网络设备和仿真平台

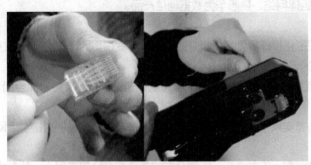

图 1-20 线缆放入水晶头压紧

图 1-21 线缆测试

① 直连法（两端均采用 T568B 标准的接法）实验结果：指示灯亮的次序如下。

1-1 灯亮，2-2 灯亮，3-3 灯亮，4-4 灯亮，5-5 灯亮，6-6 灯亮，7-7 灯亮，8-8 灯亮。

② 交叉法（一端采用 T568A 标准的接法，另一端采用 T568B 标准的接法）实验结果：指示灯亮的次序如下。

1-3 灯亮，2-6 灯亮，3-1 灯亮，4-4 灯亮，5-5 灯亮，6-2 灯亮，7-7 灯亮，8-8 灯亮。

【任务拓展】

准备多台计算机，多台集线器，使用制作好的直连线缆和交叉线缆连接计算机进行测试。

 注意　利用双绞线进行设备之间的连接时，要注意不同设备间所使用的线缆是不同的。两台计算机之间的连接使用交叉线缆。计算机与交换机/集线器之间的连接使用直连线缆。集线器与集线器的级联可以采取两种方法（交换机与交换机的级联同集线器间的级联）：一台集线器的级联口与另一台集线器的普通口连接时采用直连线缆；两台集线器的普通口连接（即级联）时采用交叉线缆。

任务 1.2　使用 eNSP 搭建一个简单的端到端网络

【知识准备】

1.2.1　eNSP 与相关网络软件概述

1. eNSP 概述

本书所有的任务实施都是在华为技术有限公司（简称华为）网络仿真工具平台 eNSP（enterprise Network Simulation Platform）上进行的。eNSP 是一款由华为提供的免费的、可扩展的、支持图形化操作的网络仿真平台，主要对企业网路由器、交换机等设备进行软件仿真，完美呈现真实设备实景，支

13

持大型网络模拟,让广大华为技术爱好者有机会在没有真实设备的情况下学习网络技术并进行模拟演练。eNSP 中融合了微型计算机系统(Microcomputer System,MCS)、客户端(Client)、服务器(Server)、无线终端等,可以完美支持组播测试、超文本传送协议(Hypertext Transfer Protocol,HTTP)测试、应用服务测试、无线测试等环境的搭建。eNSP 构建了易用的、可扩展的图形化网络仿真平台,使学员和工程师能够很方便地学习网络知识、模拟组建网络、熟悉华为数据通信产品。

eNSP 的主要特点如下。

(1)图形化操作。eNSP 提供便捷的图形化操作界面,使复杂的组网操作变得更简单,可以直观感受设备形态,并且支持一键获取华为官网的设备资料。

(2)高仿真度。eNSP 按照真实设备支持特性情况进行模拟,模拟的设备形态多,支持功能全面,模拟程度高。

(3)可与真实设备对接。eNSP 支持与真实网卡的绑定,以实现模拟设备与真实设备的对接,组网更灵活。

(4)分布式部署。eNSP 不仅支持单机部署,还支持 Server 端分布式部署在多台服务器上。分布式部署环境下能够支持更多设备组成复杂的大型网络。

2. WinPcap 概述

WinPcap 是 Windows 操作系统中一款免费、公共的网络访问软件,可用于 Windows 操作系统中的直接网络编程,其主要功能是独立于主机协议发送和接收原始数据报。WinPcap 不能阻塞、过滤或控制其他应用程序数据报的收发,它只是监听共享网络上传送的数据报。很多工具软件使用 WinPcap 用于网络分析、故障排除、网络监控等。WinPcap 特别适用于以下这几个经典领域:网络及协议分析、网络监控、通信日志记录、流量发生器、网络入侵检测系统(Network Intrusion Detection System,NIDS)、网络扫描、安全工具。

3. Wireshark 概述

Wireshark 是一款网络封包分析软件,其功能是撷取网络封包,尽可能显示出详细的网络封包资料。Wireshark 使用 WinPcap 作为接口,直接与网卡进行数据报文交换。eNSP 使用 Wireshark 进行抓包和网络分析。Wireshark 的主要用途如下。

(1)网络管理员使用 Wireshark 检测网络问题。

(2)网络安全工程师使用 Wireshark 检查信息安全相关问题。

(3)开发者使用 Wireshark 对新的通信协议进行调试。

(4)普通用户使用 Wireshark 学习网络协议相关知识。

4. VirtualBox 概述

VirtualBox 是一款开源虚拟机软件。VirtualBox 支持虚拟机克隆、Direct3D 等,可以为用户带来快速、便捷的使用体验。在安装 eNSP 时,网络中通常要引入服务器以提供相应的服务,eNSP 可以模拟路由器、交换机等网络设备,但是无法模拟服务器。为此,在 eNSP 网络中引入 VirtualBox 虚拟机,从而实现相关的功能。在加载 eNSP 设备时,VirtualBox 会自动在后台运行,无须人工激活。

1.2.2 eNSP 的安装和启动

1. eNSP 的安装

这里不对 WinPcap、Wireshark、VirtualBox 的安装进行介绍,感兴趣的读者可以参考官方资料进行安装和使用。安装 eNSP 时需注意以下事项。

V1-2 eNSP 的安装和启动

① eNSP 只支持在 Windows 操作系统中上进行安装。
② eNSP 可以通过裸机进行安装，也可以通过虚拟机进行安装。
③ 安装 eNSP 前，需要先在系统中安装好 WinPcap、Wireshark、VirtualBox，eNSP 的正常使用依赖于这 3 款软件，如图 1-22 所示。

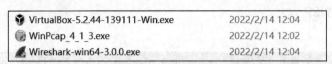

图 1-22　安装 eNSP 前需安装好的 3 款软件

VirtualBox 安装的注意事项如下。
① 注意 VirtualBox 的安装版本。
② 安装 VirtualBox 时，请不要安装在带中文的路径下，否则可能导致仿真设备无法启动。

安装 eNSP 的具体步骤如下。

（1）在华为官网上下载最新版的 eNSP。在百度首页搜索"华为官网"，如图 1-23 所示。打开华为官网，单击"解决方案&服务"的"服务"中的"培训与认证"链接，如图 1-24 所示，鼠标向下滚动找到"工具专区"中的"eNSP"，如图 1-25 所示。

图 1-23　搜索"华为官网"

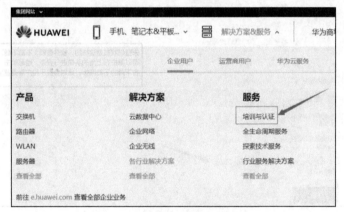

图 1-24　华为官网

图 1-25　"工具专区"中的"eNSP"

单击图 1-25 中箭头所指位置，进入图 1-26 所示的 eNSP 链接页面，默认未登录不可下载，注册并登录后可进入 eNSP 下载页面，如图 1-27 所示。

图 1-26 eNSP 链接页面

图 1-27 eNSP 下载页面

（2）eNSP 下载完成后，其安装包如图 1-28 所示，双击文件"eNSP V100R002C00B510 Setup.exe"，依据提示进行安装。

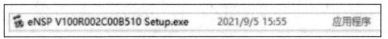

图 1-28 eNSP 安装包

（3）选择安装语言，这里选择"中文(简体)"选项，单击"确定"按钮，如图 1-29 所示。

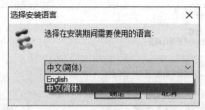

图 1-29　选择安装语言

（4）进入"欢迎使用 Enterprise Network Simulation Platform(eNSP)安装向导"界面，单击"下一步"按钮，如图 1-30 所示。

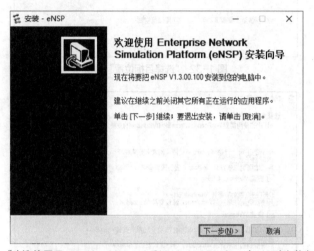

图 1-30　"欢迎使用 Enterprise Network Simulation Platform(eNSP)安装向导"界面

（5）进入"许可协议"界面，选中"我愿意接受此协议"单选按钮，如图 1-31 所示。

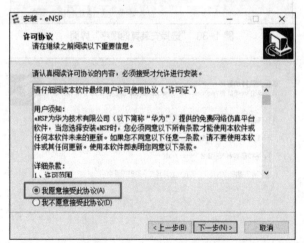

图 1-31　"许可协议"界面

（6）进入"选择目标位置"界面，单击"浏览"按钮并选择安装路径，如图 1-32 所示。

（7）进入"选择安装其他程序"界面，按提示选择安装其他程序，如图 1-33 所示。若勾选程序复选框，则软件会自动安装对应程序。如果已安装这些程序，则可以忽略，直接单击"下一步"按钮。图 1-34 所示为已安装好前面所述 3 款软件的界面。

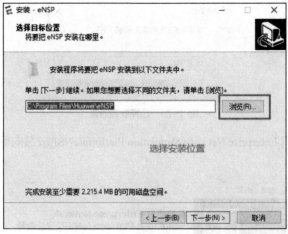

图 1-32 "选择目标位置"界面

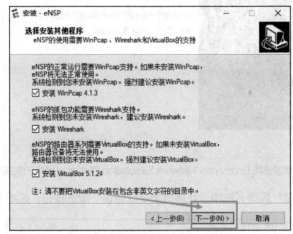

图 1-33 "选择安装其他程序"界面

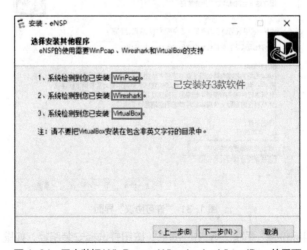

图 1-34 已安装好 WinPcap、Wireshark、VirtualBox 的界面

（8）进入"准备安装"界面，单击"安装"按钮，如图 1-35 所示。

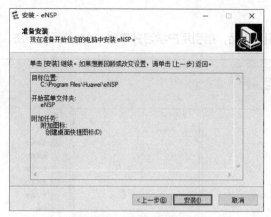

图 1-35 "准备安装"界面

(9)进入"正在完成 eNSP 安装向导"界面,单击"完成"按钮,如图 1-36 所示。

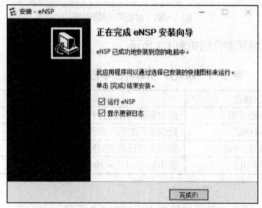

图 1-36 "正在完成 eNSP 安装向导"界面

2. eNSP 的启动

启动 eNSP,进入 eNSP 启动主界面,如图 1-37 所示。

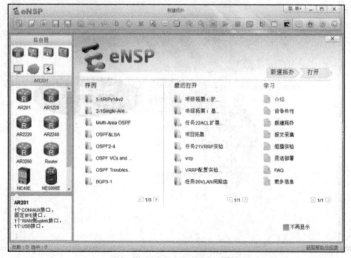

图 1-37 eNSP 启动主界面

（1）引导界面介绍

打开eNSP，将进入引导界面，指引用户学习并掌握eNSP的操作方法，如图1-38所示。

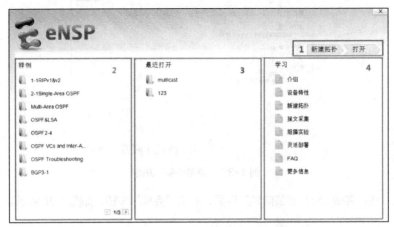

图1-38　eNSP引导界面

eNSP引导界面中各区域简要介绍如表1-1所示。

表1-1　eNSP引导界面中各区域简要介绍

序号	区域名	说明
1	快捷按钮	提供新建拓扑和打开拓扑的操作入口
2	样例区	提供常用的拓扑案例
3	最近打开区	显示最近已浏览的拓扑文件名称
4	学习区	提供学习eNSP操作方法的入口

（2）eNSP主界面介绍

eNSP主界面如图1-39所示，各区域简要介绍如表1-2所示。

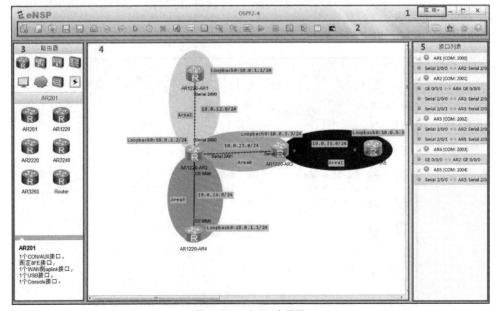

图1-39　eNSP主界面

表1-2　eNSP主界面中各区域简要介绍

序号	区域名	说明
1	主菜单	提供"文件""编辑""视图""工具""考试""帮助"等菜单
2	工具栏	提供常用的工具按钮，如"新建拓扑""打印"等
3	网络设备区	提供设备和网线，可选择到工作区中
4	工作区	在此区域中创建网络拓扑
5	设备接口区	显示拓扑中的设备和设备已连接的接口

1.2.3　VRP概述

随着越来越多的终端接入网络中，网络设备的负担也越来越重，这时网络设备可以通过华为专有的通用路由平台（Versatile Routing Platform，VRP）操作系统来提升运行效率。熟悉使用VRP进行配置和操作是对网络工程师的一项基本要求。

V1-3　VRP概述

1. VRP简介

VRP是华为从低端到高端的全系列路由器、交换机等数据通信产品的通用网络操作系统（Network Operating System，NOS），就如同微软公司的Windows操作系统之于个人计算机（Personal Computer，PC），苹果公司的iOS之于iPhone。VRP可以运行在多种硬件平台之上，并拥有一致的网络界面、用户界面和管理界面，可为用户提供灵活而丰富的应用解决方案。NOS是指运行于一定设备上的、提供网络接入及互联服务的系统软件，它主要实现系统资源的合理分配和调度、网络通信服务、用户访问控制、系统安全管理、应用服务管理等。独立、高效、稳定的专用NOS是满足网络规模不断增长、互联网技术不断发展需求，保证网络业务和网络质量的关键。

2. VRP体系结构

VRP平台以TCP/IP协议族为核心，实现了数据链路层、网络层和应用层的多种协议，在操作系统中集成了路由交换技术、服务质量（Quality of Service，QoS）技术、安全技术和IP语音技术等，并以IP转发引擎技术作为基础，为网络设备提供了出色的数据转发能力。VRP体系结构示意如图1-40所示。

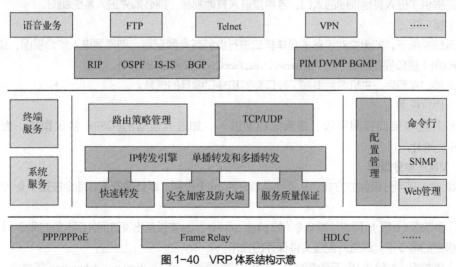

图1-40　VRP体系结构示意

（1）广域网互联：支持PPP/PPPoE、HDLC、帧中继（Frame Relay）等。

（2）IP转发引擎：主要有单播转发和多播转发，主要功能包括快速转发、安全加密及防火墙、服务质量保证等。

（3）路由协议：支持 RIP、OSPF 协议、IS-IS 协议、BGP、PIM 协议、DVMRP、BGMP 等。

（4）IP 业务：支持 ARP/Proxy ARP、NAT、DNS、DHCP 中继、VLAN、SNA、VoIP 和 VPN 等。

3. VRP 命令行视图

VRP 的命令行界面分成了若干种命令行视图，使用某个命令行时，需要先进入该命令行所在的视图。常用的命令行视图有用户视图、系统视图、接口视图和协议视图，它们之间既有联系，又有一定的区别。视图模式如图 1-41 所示。

图 1-41 视图模式

（1）用户视图

进入命令行界面后，首先进入的就是用户视图，提示符为<Huawei>。

以上提示符中，"< >"表示用户视图，"Huawei"表示设备默认的主机名。在用户视图下，用户可以查询设备的基础信息，但不能进行与业务功能相关的配置。如果需要对设备进行与业务功能相关的配置，则需要进入系统视图。

（2）系统视图

在用户视图下使用 system-view 命令，便可以进入系统视图，提示符为[Huawei]。

此时的提示符使用了方括号"[]"。系统视图下可以使用绝大部分的基础功能配置命令。另外，系统视图还提供了进入其他视图的入口。若希望进入其他视图，则必须先进入系统视图。

（3）接口视图

在系统视图下，如果要对设备的具体接口进行业务或参数配置，则需要进入接口视图，如进入设备的 GE 0/0/1 接口视图的提示符为[Huawei-interfaceGigabitEthernet0/0/1]。

进入接口视图后，主机名后追加了接口类型和接口编号的信息。

（4）协议视图

协议视图与接口视图平级，由系统视图进入，如进入设备的 OSPF 协议模式，提示符为[Huawei-ospf-1]。

4. VRP 命令级别

VRP 命令采用分级保护方式，从低到高划分为 16 个级别。默认情况下，命令按如下 0~3 级进行注册。

（1）参观级（0 级）：网络诊断命令（如 ping、tracert）、从本设备出发访问外部设备的命令（如 telnet）等，该级别的命令不允许进行配置文件保存的操作。

（2）监控级（1 级）：用于系统维护、业务故障诊断等，包括 display、debugging 等命令，该级别的命令不允许进行配置文件保存的操作。

（3）系统级（2 级）：业务配置命令，包括路由、各个网络层次的命令，这些命令用于向用户提供直接的网络服务。

（4）管理级（3级）：关系到系统基本运行、系统支撑模块的命令，这些命令对业务起支撑作用，包括文件系统命令、文件传送协议（File Transfer Protocol，FTP）命令、简易文件传送协议（Trivial File Transfer Protocol，TFTP）命令、配置文件切换命令、电源控制命令、备板控制命令、用户管理命令、级别设置命令、系统内部参数设置命令（非协议规定）等。

需要注意的是，在默认情况下，4~15级的用户权限与3级的用户权限是一致的。

5. VRP命令帮助

在命令行输入过程中，用户可以随时输入"？"以获得帮助信息，从而无须记忆大量的复杂命令。

（1）完全帮助

在任意命令行视图下，输入"？"，可获取该命令行视图下所有命令及其简单描述，示例如下。

```
<Huawei>?
User view commands:
  arp-ping              ARP-ping
  autosave              <Group> autosave command group
  backup                Backup information
  cd                    Change current directory
  clear                 <Group> clear command group
  clock                 Specify the system clock
  cls                   Clear screen
  compare               Compare configuration file
  copy                  Copy from one file to another
  debugging             <Group> debugging command group
  delete                Delete a file
  dialer                Dialer
  dir                   List files on a filesystem
  display               Display information
  factory-configuration Factory configuration
  fixdisk               Try to restory disk
  format                Format file system
  free                  Release a user terminal interface
  ftp                   Establish an FTP connection
  help                  Description of the interactive help system
  hwtacacs-user         HWTACACS user
  license               <Group> license command group
  lldp                  Link Layer Discovery Protocol
 ---- More ----
```

输入一条命令的部分关键字，后接以空格分隔的"？"，如果该位置为关键字，则可列出全部关键字及其简单描述，示例如下。

```
[Huawei]aaa ?
  abnormal-offline-record  Abnormal-offline-record
```

```
    offline-record              Offline-record
    online-fail-record          Online-fail-record
  <cr>                          Please press ENTER to execute command
```

其中,"<cr>"表示该位置没有关键字或参数,直接按 Enter 键即可执行命令。

(2)部分帮助

当用户需输入命令,但只记得命令关键字的部分字符时,可以使用命令行的部分帮助功能,获取包含该字符串的所有关键字的提示。

输入某字符串,其后紧接"?",可列出以该字符串开头的所有关键字,示例如下。

```
<Huawei>d?
  debugging   <Group> debugging command group
  delete      Delete a file
  dialer      Dialer
  dir         List files on a filesystem
  display     Display information
```

输入命令的某个关键字的前几个字母,按 Tab 键,可以显示出完整的关键字,前提是这几个字母可以唯一标识出该关键字;否则,连续按 Tab 键,可出现不同的关键字,可从中选择所需要的关键字。

(3)历史帮助

组合键 Ctrl+P 用于显示上一条历史命令。如果需要显示更早的历史命令,则可以重复使用该组合键。组合键 Ctrl+N 用于显示下一条历史命令。

简单命令的功能及其所属视图的对应关系如表 1-3 所示。

表 1-3 简单命令的功能及其所属视图的对应关系

命令	功能	视图
Ctrl+A	光标到当前命令行最前端	
Ctrl+C	停止当前命令的运行	
Ctrl+Z	回到用户视图	
Ctrl+]	终止当前连接或切换连接	
Backspace	删除光标左边第一个字符	
←或 Ctrl+B	光标左移	
→或 Ctrl+F	光标右移	
Tab	补全不完整命令	
sysname	配置设备名称	系统视图
clock timezone	配置所在时区	用户视图
clock datetime	配置当前时间和日期	用户视图
clock daylight-saving-time	配置夏令时	用户视图
display clock	查看时钟	用户视图
header login information "标题"	配置用户登录前的标题	系统视图
header shell information "标题"	配置用户登录后的标题	系统视图
quit	退出当前视图(只能一级一级退出)	
return(Ctrl+Z)	直接退出到用户视图	
reset saved-configuration	清除设备的所有配置(恢复出厂设置)	
clear configuration 接口名称	清除该接口的所有配置	

6. 参数基本配置

（1）配置设备名称

命令行界面中的尖括号"<>"或方括号"[]"中包含设备的名称，也称为设备主机名。默认情况下，设备名称为"Huawei"。为了更好地区分不同的设备，通常需要修改设备名称。我们可以通过命令 sysname host-name 来对设备名称进行修改。其中，sysname 是命令关键字；host-name 为参数，表示希望设置的设备名称。配置设备名称的示例如图1-42所示。

```
[Huawei]sysname R1
May  5 2020 13:24:56-08:00 Huawei DS/4/DATASYNC_CFGCHANGE:OID 1.3.6.1.4.1.2011.5
.25.191.3.1 configurations have been changed. The current change number is 3, th
e change loop count is 0, and the maximum number of records is 4095.
[R1]
```

图1-42　配置设备名称的示例

（2）配置设备系统时钟

华为设备出厂时默认采用了协调世界时（Universal Time Coordinated，UTC），但没有配置时区，所以在配置设备系统时钟前，需要了解设备所在的时区。

配置时区的命令为 clock timezone time-zone-name {add|minus}。其中，time-zone-name 为用户定义的时区名，用于标识配置的时区；根据偏移方向选择 add 和 minus，正向偏移（UTC 加上偏移量为当地时间）选择 add，负向偏移（UTC 减去偏移量为当地时间）选择 minus。假设设备位于北京时区，则相应的配置命令如下。

```
clock timezone BJ add 08:00
```

配置好时区后，就可以配置设备当前的日期和时间了。华为设备仅支持 24 小时制，使用的命令为 clock datetime HH:MM:SS YYYY-MM-DD。其中，HH:MM:SS 为配置的时间；YYYY-MM-DD 为配置的日期。假设当前的日期是 2023 年 10 月 4 日，时间是 02:06:00，则相应的配置命令如下。

```
clock datetime 02:06:00 2023-10-04
```

读者对配置设备系统时钟的方法有所了解即可，在真实环境中使用频率较低。

（3）配置设备 IP 地址

用户可以通过不同的方式登录到设备命令行界面，包括 Console 口登录、MiniUSB 口登录和 Telnet 登录。首次登录新设备时，因为新设备为空配置设备，所以只能通过 Console 口或 MiniUSB 口登录。首次登录到新设备后，便可以给设备配置一个 IP 地址，并启用 Telnet 功能。

IP 地址是针对设备接口的配置，通常一个接口配置一个 IP 地址。配置接口 IP 地址的命令为 ip address ip-address{masl^mask-length}。其中，ip address 是命令关键字；ip-address 为希望配置的 IP 地址；mask 表示点分十进制方式的子网掩码；mask-length 表示长度方式的子网掩码，即掩码中二进制数 1 的个数。

假设设备 R 的管理接口为 GE0/0/0，分配的 IP 地址为 10.1.1.100，子网掩码为 255.255.255.0（24），则相应的配置命令如图1-43所示。

图1-43　配置设备 IP 地址的示例

1.2.4　网络设备系统登录管理

网络设备系统登录管理主要有本地登录管理和远程登录管理两种模式。

（1）本地登录管理模式主要使用 Console 线路登录管理。
（2）远程登录管理模式使用虚拟终端（Virtual Teletype，VTY）线路进行 Telnet 登录管理。
　　VTY 线路可以同时允许多个用户登录，而 Console 线路只允许一个用户登录。登录管理的验证方式有 NONE（无验证）、PASSWORD（密码）、AAA 认证这 3 种。
　　VTY 登录模式有 Telnet 登录模式和安全外壳（Secure Shell，SSH）登录模式。

【任务实施】

实验 2　搭建一个简单的端到端网络

实验目的：
（1）熟悉 eNSP 的设备类型。
（2）了解设备接口类型，熟悉简单设备的拓扑搭建。
（3）熟悉设备的启动及参数的设置。
（4）了解 Wireshark 在本地连接口抓包的设置。
（5）熟悉 eNSP 中 Wireshark 的数据抓包过程，分析捕获的数据包的流量。

V1-4　搭建一个简单的端到端网络

　　本实验模拟办公室的两台主机通过交换机连接在一起，要求两台主机的 IP 地址配置在同一网段，通过交换机实现只有两个节点的计算机网络，测试主机间的连通性，并在交换机接口上抓包测试、分析。

1. 搭建简单网络拓扑

　　使用 eNSP 搭建的简单网络拓扑示意如图 1-44 所示。启动 eNSP，新建拓扑文件，在工作区中创建一台交换机 S5700，创建两台终端 PC（PC1 和 PC2）。通过设备连线连接终端和交换机的相应接口，详细步骤如下。

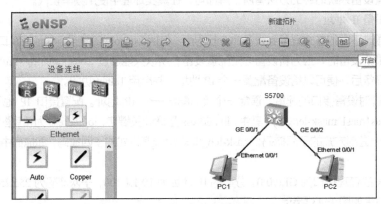

图 1-44　使用 eNSP 搭建的简单网络拓扑示意

　　（1）单击该窗口左上角的"新建拓扑"图标，创建一个新的实验场景，可以在进入的空白界面中搭建网络拓扑，练习组网并分析网络行为。本实验中需要使用两台终端建立一个简单的端到端网络。
　　（2）在左侧面板顶部，单击"终端"图标，在显示的终端设备中，选中"PC"图标，把图标拖动到空白界面中。使用相同步骤，再拖动一个"PC"图标到空白界面中。PC 设备模拟的是终端主机，可以再现真实的操作场景。
　　（3）在左侧面板顶部，单击"交换机"图标，在显示的终端设备中，选中"S5700"图标，把图标拖动到空白界面中，将其作为 PC 设备之间的桥接。

（4）在左侧面板顶部，单击"设备连线"图标，在显示的媒介中，选择"Copper"图标。单击图标后，光标代表一个连接器。单击客户端 PC1 设备的 Ethernet 0/0/1 接口与交换机设备，会显示对应该模拟设备包含的所有接口，单击 GE 0/0/1 接口即可将其连接起来。同理，单击客户端 PC2 设备的 Ethernet 0/0/1 接口，将其与交换机设备的 GE 0/0/2 接口连接起来。

（5）在图 1-44 所示的已建立的端到端网络中，连线的两端显示的是两个圆点（此时显示为红色），表示该连线连接的两个端口都处于 Down 状态，如果为绿色圆点，则表示该连线连接的两个端口都处于 Up 状态。

（6）启动工作区的设备。右键单击设备，选择"启动"选项；也可以在工作区中用鼠标选定一个区域，单击工具栏中的▶按钮，批量启动该区域的设备。

（7）单击工具栏中的按钮，输入名称，并保存拓扑。

2．数据实验

（1）开启 eNSP 客户端

单击工具栏中的按钮，加载新建的拓扑。单击工具栏中的▶按钮，启动全部设备。配置两台 PC 的 IP 地址和子网掩码，参数列表拓扑示意如图 1-45 所示。在工作区中双击 PC1，在"基础配置"界面中，将 IP 地址配置为"192.168.1.1"，子网掩码配置为"255.255.255.0"，如图 1-46 所示。按照相同的方法，配置 PC2 的 IP 地址和子网掩码分别为"192.168.1.2"和"255.255.255.0"。

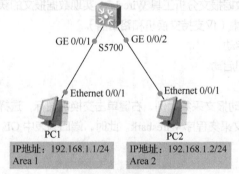

图 1-45　参数列表拓扑示意

图 1-46　PC1 基础配置参数信息

（2）报文测验

双击 PC1，打开命令行窗口，输入"ping 192.168.1.2"，表示请求连接 PC2，可以测试 PC1、PC2 是否在同一个网络。由图 1-47 可知，发送的 5 个报文都已成功，丢失率为 0.00%，说明两台 PC 已连接好网络，可以实现相互通信。

图 1-47 PC 间测试连通性结果

eNSP 利用第三方网络数据报文分析工具 Wireshark 实现数据报文的获取与分析。eNSP 提供以下多种便捷操作入口启动报文采集（仅支持交换机和路由器）。

① 在设备上指定端口启动。
② 单击工具栏中的按钮启动。
③ 接口视图链路启动。

以在设备上指定端口启动报文采集为例，右键单击交换机图标，选择"数据抓包"选项，再选择端口（见图 1-48），启动报文采集程序 Wireshark。此时，端口列表中 GE 0/0/2 的指示灯变为蓝色。

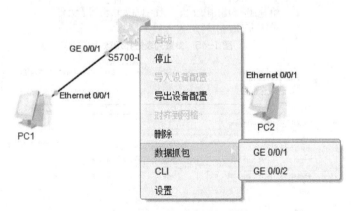

图 1-48 端口 GE 0/0/2 抓包示意图

从采集报文的 PC 向另一台 PC 发送 ping 报文，在 Wireshark 上查看获取的 ping 报文，如图 1-49 所示。如需监控更多接口，则重复上述步骤，选择不同接口即可，Wireshark 将会为每个接口激活不同实例来捕获数据包。根据被监控设备的状态，Wireshark 可捕获选中接口上产生的所有流量，生成抓包结果。

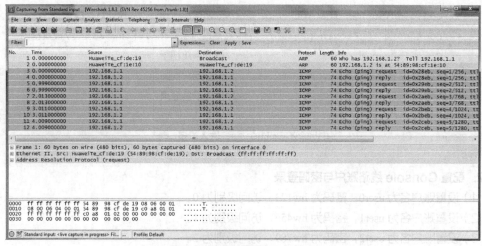

图 1-49 抓包捕获的 ping 报文

实验 3 配置本地登录管理与远程登录管理

实验目的：
（1）熟悉本地登录管理与远程登录管理。
（2）掌握 Console 线路密码配置。
（3）熟悉访问级别权限等级设置。
（4）熟悉 VTY 线路的配置。
（5）掌握 IP 地址的配置方法。
（6）熟悉 Telnet 登录管理的配置。
（7）熟悉 SSH 登录管理的配置。

V1-5 配置本地登录管理

V1-6 配置远程登录管理

假设某单位网络出现故障，重新采购了一台新的路由器，路由器到货拆箱后，怎样完成路由器的基本配置呢？由于路由器本身没有配备显示设备和输入设备，无法直接显示运行的参数和输入命令，因此配置需要借助于有输入/输出设备的计算机，又因设备刚刚拆箱，没有配置任务网络服务，所以适合选择 Console 口进行登录。通过 Console 口登录后，完成路由器的其他基本配置。

V1-7 配置 SSH 登录管理

1. 配置 Console 线路简单密码登录

图 1-50 所示为 Console 线路配置的拓扑示意，设置路由器 R1 登录密码为 hw123。本实验使用模拟器完成配置 Console 口登录，这里 PC1 不是真机，所以只在路由器 R1 上进行配置即可。具体代码如下。

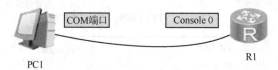

图 1-50 Console 线路配置的拓扑示意

```
<Huawei>system-view                    //从用户视图进入系统视图
[Huawei]sysname R1                     //修改主机名为 R1
[R1]user-interface Console 0           //进入 Console 线路
```

```
[R1-ui-Console0]authentication-mode password    //选择使用密码认证模式
Please configure the login password (maximum length 16):hw123    //设置密码
[R1-ui-Console0]quit                            //返回系统视图
[R1]quit                                        //返回用户视图
<R1>quit                                        //退出设备登录
Configuration Console exit, please press any key to log onLogin authentication
PasSWord:           //在此输入密码hw123,进入设备
<R1>
```

2. 配置 Console 线路账户与密码登录

（1）设置账户名为 user0，密码为 hw123，访问级别为 0。
（2）设置账户名为 user1，密码为 hw456，访问级别为 1。
（3）设置账户名为 user2，密码为 hw789，访问级别为 2。
配置 AAA 模式本地管理密码，代码如下。

```
[R1]user-interface Console 0                              //进入 Console 线路
[R1-ui-Console0]authentication-mode AAA                   //选择使用 AAA 认证模式
[R1-ui-Console0]quit                                      //返回系统视图
[R1]AAA                                                   //进入 AAA 认证模式
[R1-AAA]local-user user0 password cipher huawei123        //创建账户 user0 并设置其密码
[R1-AAA]local-user user1 password cipher huawei456        //创建账户 user1 并设置其密码
[R1-AAA]local-user user2 password cipher huawei789        //创建账户 user2 并设置其密码
[R1-AAA]local-user user0 privilege level 0                //分配访问级别 0
[R1-AAA]local-user user1 privilege level 1                //分配访问级别 1
[R1-AAA]local-user user2 privilege level 2                //分配访问级别 2
[R1-AAA]quit                                              //返回系统视图
[R1]quit                                                  //返回用户视图
<R1>quit                                                  //退出设备登录
```

退出设备登录，并通过不同的用户进入设备。设备等级不同，访问的权限也有限制。

3. 配置 VTY 线路登录

新路由器经过 Console 口登录后，可以进行 IP 地址及相关的参数配置。我们可以利用远程终端设备，连接此路由器进行远程配置。VTY 线路配置拓扑示意如图 1-51 所示。拓扑中包含两台路由器 R1、R2，其中，R1 为登录设备（远程终端设备），R2 为被登录设备。

图 1-51 VTY 线路配置拓扑示意

（1）配置端口 IP 地址，代码如下。

```
[R1]int g0/0/0                                            //进入 GE 0/0/0 端口
[R1-GigabitEthernet0/0/0]ip address 12.1.1.1 24           //配置 IP 地址
```

```
[R1-GigabitEthernet0/0/0]quit                    //返回系统视图

[R2]int GigabitEthernet 0/0/0
[R2-GigabitEthernet0/0/0]ip address 12.1.1.2 24
[R2-GigabitEthernet0/0/0]quit
[R2]
```

(2) 查看端口 IP 地址，代码如下。

```
[R1]display ip interface brief          //查看R1端口的IP地址简要信息
*down: administratively down
^down: standby
(l): loopback
(s): spoofing
The number of interface that is UP in Physical is 2
The number of interface that is DOWN in Physical is 2
The number of interface that is UP in Protocol is 2
The number of interface that is DOWN in Protocol is 2

Interface                  IP Address/Mask      Physical    Protocol
GigabitEthernet0/0/0       12.1.1.1/24          up          up
GigabitEthernet0/0/1       unassigned           down        down
GigabitEthernet0/0/2       unassigned           down        down
NULL0                      unassigned           up          up(s)
[R1]

[R2]display ip interface brief          //查看R2端口的IP地址简要信息
*down: administratively down
^down: standby
(l): loopback
(s): spoofing
The number of interface that is UP in Physical is 2
The number of interface that is DOWN in Physical is 2
The number of interface that is UP in Protocol is 2
The number of interface that is DOWN in Protocol is 2

Interface                  IP Address/Mask      Physical    Protocol
GigabitEthernet0/0/0       12.1.1.2/24          up          up
GigabitEthernet0/0/1       unassigned           down        down
GigabitEthernet0/0/2       unassigned           down        down
NULL0                      unassigned           up          up(s)
[R2]
```

（3）测试连通性。通过 ping 命令测试 R1 与 R2 的连通性，代码如下。

```
<R1>ping 12.1.1.2
  PING 12.1.1.2: 56  data bytes, press CTRL_C to break
    Reply from 12.1.1.2: bytes=56 Sequence=1 ttl=255 time=80 ms
    Reply from 12.1.1.2: bytes=56 Sequence=2 ttl=255 time=20 ms
    Reply from 12.1.1.2: bytes=56 Sequence=3 ttl=255 time=20 ms
    Reply from 12.1.1.2: bytes=56 Sequence=4 ttl=255 time=20 ms
    Reply from 12.1.1.2: bytes=56 Sequence=5 ttl=255 time=20 ms
  --- 12.1.1.2 ping statistics ---
    5 packet(s) transmitted
    5 packet(s) received
  0.00% packet loss
    round-trip min/avg/max = 20/32/80 ms
```

（4）配置 Telnet。在被登录设备 R2 上配置 Telnet，代码如下。

```
[R2]user-interface VTY 0 4                              //进入 VTY 线路模式
[R2-ui-VTY0-4]authentication-mode AAA                   //选择使用 AAA 模式的认证
[R2-ui-VTY0-4]quit                                      //返回系统视图
[R2]AAA                                                 //进入 AAA 认证模式
[R2-AAA]local-user user-r1 password cipher huawei123    //创建账户 user-r1
[R2-AAA]local-user user-r1 privilege level 2            //设置访问级别为 2
[R2-AAA]local-user user-r1 service-type Telnet          //定义用户的服务类别为 Telnet
[R2-AAA]quit                                            //返回系统视图
```

（5）在 R1 上使用 telnet 命令登录 R2，代码如下。

```
<R1>telnet 12.1.1.2              //在用户模式下进行远程登录
  Press CTRL_] to quit telnet mode
  Trying 12.1.1.2 ...
  Connected to 12.1.1.2 ...

Login authentication
Username:user-r1                 //输入账户名
Password:                        //输入密码
<R2>
<R2>system-view                  //出现 R2 模式，登录成功
[R2]sysname R22                  //修改 R2 设备的名称为 R22
[R22]
```

（6）配置 SSH 服务。在被登录设备 R2 上配置 SSH 服务，代码如下。

```
[R2]stelnet server enable         //启用 SSH 协议
Info: Succeeded in starting the STelnet server.
[R2]rsa local-key-pair create    //创建加密报文的密钥对
The key name will be: Host
```

```
% RSA keys defined for Host already exist.
Confirm to replace them? (y/n)[n]:y
The range of public key size is (512 ~ 2048).
NOTES: If the key modulus is greater than 512,
       It will take a few minutes.
Input the bits in the modulus[default = 512]:1024
Generating keys...
...........++++++
......++++++
...++++++++
............++++++++

[R2]aaa                                      //进入 AAA 认证模式
[R2-aaa]local-user user-ssh password cipher hw123    //创建账户 user-ssh
Info: Add a new user.
[R2-aaa]local-user user-ssh privilege level 2    //设置访问级别为 2
[R2-aaa]local-user user-ssh service-type ssh     //定义用户的服务类别为 SSH
[R2-aaa]qu
[R2]user-interface vty 0 4                   //进入 VTY 线路模式
[R2-ui-vty0-4]authentication-mode aaa        //选择使用 AAA 模式的认证
[R2-ui-vty0-4]protocol inbound ssh           //启用 VTY 线路的 SSH 访问功能
[R2-ui-vty0-4]quit
[R2]ssh user user-ssh authentication-type all    //定义 SSH 用户的认证模式
 Authentication type setted, and will be in effect next time
```

（7）SSH 登录。在登录设备 R1 上通过 SSH 登录 R2，且修改 R2 的主机名为 R22，代码如下。

```
<R1>sys
Enter system view, return user view with Ctrl+Z.
[R1]ssh client first-time enable             //启用 SSH
[R1]stelnet 12.1.1.2                         //在系统视图下进行远程登录
Please input the username:user-ssh           //输入账户名 user-ssh
Trying 12.1.1.2 ...
Press CTRL+K to abort
Connected to 12.1.1.2 ...
The server is not authenticated. Continue to access it? (y/n)[n]:y 是否继续访问？y
Sep  7 2021 08:54:30-08:00 R1 %%01SSH/4/CONTINUE_KEYEXCHANGE(l)[0]:The server ha
d not been authenticated in the process of exchanging keys. When deciding whether
 to continue, the user chose Y.
[R1]
Save the server's public key? (y/n)[n]:y
The server's public key will be saved with the name 12.1.1.2. Please wait...
```

```
Sep  7 2021 08:54:38-08:00 R1 %%01SSH/4/SAVE_PUBLICKEY(l)[1]:When deciding wheth
er to save the server's public key 12.1.1.2, the user chose Y.
[R1]
Enter password:                                    //输入密码hw123
<R2>                                               //SSH登录成功
<R2>system-view
Enter system view, return user view with Ctrl+Z.
[R2]sys
[R2]sysname R22                                    //修改R2的名称为R22
[R22]
```

【任务拓展】

（1）使用 eNSP 搭建一个简单的校园网络拓扑，配置所有 PC 的基础参数，通过 ping 命令测试连通情况，通过 Wireshark 捕获 SW3 交换机上的 GE 0/0/1 端口信息并进行分析。简单的校园网络拓扑示意如图 1-52 所示。

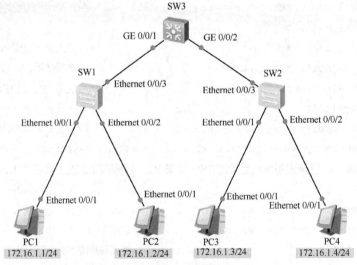

图 1-52　简单的校园网络拓扑示意

（2）根据图 1-53 所示的设备管理配置，完成以下任务。

① 通过 Console 口连接路由器。
② 配置设备 Console 密码、主机名、时间、时区等。
③ 配置端口地址与 Telnet、SSH 登录。
④ 查看、保存、清空、重启路由器。

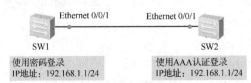

图 1-53　设备管理配置

模块 2
交换机组网技术

随着局域网中计算机、交换机等网络设备数量的不断增加，网络流量随之增大，而网络中的广播风暴、安全性等问题也随之而来。如果能很好地应用交换机组网技术，如交换机、VLAN、链路聚合等技术，就能很大程度地解决网络环路，隔离广播风暴，增加网络带宽，实现链路负载均衡，提高网络可靠性。本模块主要介绍交换机组网技术，包括配置 STP、VLAN 划分、配置链路聚合技术等。

【学习目标】

知识目标
（1）理解 VLAN 的划分作用及应用场景。
（2）理解 Access 端口、Trunk 端口、Hybrid 端口的作用。
（3）理解交换机的工作原理。
（4）理解根桥、根端口、指定端口的概念。
（5）理解链路聚合的作用。

技能目标
（1）掌握 VLAN 的 Access 端口、Trunk 端口、Hybrid 端口的配置方法。
（2）掌握根桥、根端口、指定端口角色的查看方法。
（3）能熟练更换交换机根桥，掌握根端口的优先级参数的设置方法。
（4）熟悉手动负载均衡模式和静态 LACP 模式。

素质目标
（1）增强学生灵活运用所学知识解决实际问题的能力。
（2）培养学生虚拟网络空间的想象力。

任务 2.1 配置 STP

【知识准备】

2.1.1 STP

1. 网络环路形成

在规模较大的局域网中，时常会遇到网络通道被严重堵塞的现象，造成这种故障现象的原因有很多，如网络遭遇病毒攻击、网络设备发生硬件损坏、网络端口出现传输瓶颈等。从网络堵塞现象发生的统计概率来看，网络中被改动或发生过变化的位置最容易发生故障现象，因为频繁改动网络时很容易引发网络环路，而由网络环路引起的网络堵塞现象常常具有较强的隐蔽性，不利于故障现象的高效排除。

2. STP 概述

生成树协议（Spanning Tree Protocol，STP）是在 IEEE 802.1d 中定义的数据链路层协议，可应用于计算机网络中树形拓扑结构的建立，主要作用是防止网络中冗余链路形成环路。STP 通过协议收敛使交换网络中任意两台设备之间有且只有一条可行路径，这样既可以避免单点故障问题，又可以解决广播风暴问题。对于任何一个商用网络来说，冗余都是一个必须考虑的问题。简单地说，网络的冗余设计主要包含关键设备冗余和关键链路冗余。

以图 2-1 所示的网络为例，如果 SW1 与 SW3 之间的链路发生故障，或者 SW1 发生故障，那么 PC 就无法到达外部网络了，因此该网络的冗余性较差。

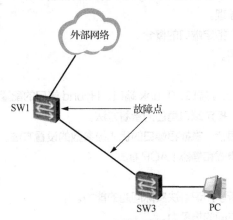

图 2-1 冗余性较差的网络

在典型的园区网络中，通常会部署冗余的设备或者冗余的链路，从而使业务流量在故障发生时能够通过冗余的设备及链路进行转发。如图 2-2 所示，网络中增加了一台交换机 SW2，SW3 通过以太网链路分别连接 SW1 和 SW2。如此一来，PC 到达外部网络的路径就拥有了冗余性。但是在这种组网中，SW1、SW2、SW3 及其链路就构成了一个二层环路。二层环路会引发广播风暴、设备 MAC 地址表紊乱，以及 MAC 地址漂移等一系列问题，严重时由环路引发的广播风暴更有可能耗尽链路带宽，或者使设备的 CPU 利用率急剧攀升并导致其无法正常工作，最终造成网络瘫痪。当然，除了前面所述的场景外，在实际的网络中还有不少二层环路是人为的疏忽导致的，如误接网络线缆等。

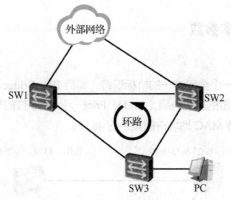

图 2-2 二层环路网络

如何在保证网络冗余性的情况下,消除二层环路呢?以太网二层环路问题的最典型的解决方案就是 STP。简单地说,当网络的交换机中部署 STP 之后,交换机之间便会开始交互相关协议报文,并在网络中进行一系列计算,经计算得到一个无环的网络拓扑。当网络中存在环路时,STP 会对网络中的一个或多个接口进行阻塞,从而打破二层环路。

如图 2-3 所示,被 STP 阻塞的接口不能再转发数据,这样网络中的二层环路问题便迎刃而解。

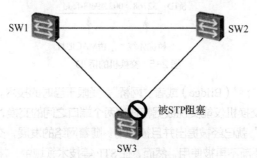

图 2-3 STP 阻塞特定端口

在此之后,STP 依然会监视网络的拓扑状况,当网络拓扑发生变更时,它能够及时感知并动态地调整被阻塞接口,这个过程无须人工干预。如图 2-4 所示,当 SW1 与 SW3 之间的链路发生故障时,STP 能感知到变化的发生,并且将原先被阻塞的接口切换到转发状态,这样一来,SW3 的上行流量可以从右侧的链路进行转发。因此,STP 不仅可以用于在网络中解决二层环路问题,还可以保证网络的冗余性。

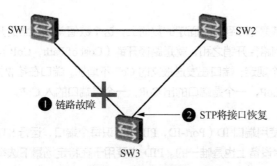

图 2-4 动态恢复链路

2.1.2 STP 的基本参数

1. 桥 ID

运行 STP 的交换机都有一个全网唯一的桥标识符，即桥 ID（Bridge-ID，BID）。如图 2-5 所示，桥 ID 一共有 64 位，包含 16 位的桥优先级（Bridge Priority）和 48 位的桥 MAC 地址，其中，桥优先级占据桥 ID 的高 16 位，而桥 MAC 地址占据其余的 48 位。

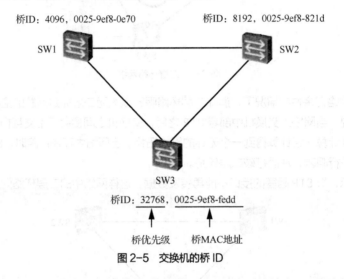

图 2-5 交换机的桥 ID

早期的交换机被称为"桥"（Bridge）或者"网桥"。受限于当时的技术，早期交换机的端口数量很少，通常只有两个端口，交换机仅能实现数据帧在这两个端口之间的交换，这也是"桥"这一称呼的由来。STP 在"网桥时代"就已经被提出并且被应用。随着网络的发展，交换机能够支持的端口数量越来越多，因此上述称呼逐渐不再被使用，然而，在 STP 等技术领域中，"桥"或"网桥"的称呼却一直被沿用下来，直至今日我们在 STP 中依然会用它们来称呼交换机。

2. 根桥

网络中拥有最小桥 ID 的交换机将成为根桥（Root Bridge，RB）。在比较桥 ID 时，首先比较的是桥优先级，桥优先级的值最小的交换机将成为根桥，如果桥优先级相等，那么桥 MAC 地址最小的交换机将成为根桥。如图 2-5 所示，交换机 SW1 为该网络中的根桥。华为交换机默认的桥优先级为 32768，桥优先级的取值是 0~61440，并且必须为 4096 的倍数，如 0、4096、8192 等。

3. 开销

STP 将不同的链路带宽对应到一个固定的数值上，这个数值就是链路的开销（Cost），从一台交换机的一个端口到达根桥的路径开销之和，就是路径开销（Cost of Path，CoP）。

在计算 CoP 时有一个规定：端口在发送报文时 CoP 不增加，端口在接收报文时 CoP 才增加，换句话说，一个端口有两个 CoP，一个是端口的出 CoP，一个是端口的入 CoP。

4. 端口 ID

运行 STP 的交换机使用端口 ID（Port-ID，PID）标识每个端口，运行 STP 的设备上的每个端口都有一个 PID，PID 在一台设备上也是唯一的。PID 主要用于在特定场景下选举指定端口，其长度为 16 位，由两部分组成，其中高 4 位是端口优先级，低 12 位是端口编号。以华为 S5700 交换机为例，默认端口优先级为 128，优先级的取值是 0~240，并且必须是 16 的倍数，如 0、16、32 等。

5. 桥协议数据单元

桥协议数据单元（Bridge Protocol Data Unit，BPDU）是 STP 定义的数据包，是在 IEEE 802.1d 中定义的，可以用来消除桥回路。网桥用 BPDU 来相互通信，并且动态选择根桥。在一个 STP 的环境中，桥不会立即实现转发功能，它们必须首先选择一个桥为根桥，然后建立一个指定路径。

2.1.3　STP 的实施过程

STP 的基本原理如下：在具有物理环路的交换网络中，交换机通过运行 STP 自动生成没有环路的网络拓扑。STP 的任务是找到网络中的所有链路，并关闭所有冗余的链路，这样就可以防止网络环路的产生。为了达到这个目的，STP 首先需要选举一个根桥，由根桥负责决定网络拓扑。所有交换机同意将某台交换机选举为根桥时，就必须为其余的交换机选定唯一的根端口，还必须为交换机之间的每一条链路两端连接的端口选定一个指定端口。既不是根端口又不是指定端口的就成为备用端口，它不转发计算机通信帧，从而阻断环路。

STP 的实施过程分为以下 4 个步骤。

1. 根桥选举

在 STP 运行之初，所有交换机都会认为自己是网络中的根桥。此后交换机通过所有端口向外发送自己的 BPDU。BPDU 中包含这台交换机的 BID 信息。交换机间获取到彼此的 BPDU 后，通过比较邻居交换机和自己的 BID 大小，从而判断真正的根桥是谁。根桥选举是"比小"，谁的 BID 更小，谁就是根桥。在一个网络中拥有最小 BID 的交换机将变成根桥，全部的 STP 网络中只有一台根桥。如图 2-6 所示，SW1、SW2 及 SW3 的桥优先级都是 32768，因此桥 MAC 地址最小的 SW1 成为网络中的根桥。

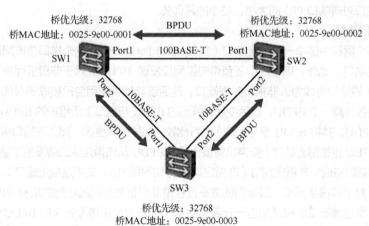

图 2-6　根桥选举

注意，根桥的地位是具有可抢占性的。以图 2-6 为例，在 STP 完成收敛后，如果网络中接入了一台新的交换机，且这台新的交换机的优先级为 4096，比现有根桥 SW1 的优先级更高，那么这台新的交换机将成为网络中的新根桥。与此同时，STP 将重新收敛、重新计算网络拓扑，在这个过程中有可能引发网络震荡，从而对业务流量的正常转发造成影响，由此可见根桥的稳定性多么重要。

2. 根端口选举

在一个交换网络中，除了根桥之外的其他交换机都是非根桥，STP 将为每个非根桥选举一个根端口。根端口实际上是非根桥上所有端口中收到最优 BPDU 的端口，即交换机在 STP 上"朝向"根桥的端口。非根桥可能会有一个或多个端口接入同一个交换网络，STP 将在这些端口之中选举出一个（且只会选举出一个）根端口。

在 STP 收敛完成之后，根桥依然会周期性地向网络中发送 BPDU，而非根桥则会周期性地在自己的根端口上收到 BPDU，并沿着 STP 向下游转发。如图 2-7 所示，SW2 及 SW3 均为非根桥，以 SW3 为例，在 STP 收敛过程中，它在自己的 Port1 及 Port2 端口上都会收到 BPDU，SW3 会对这两个 BPDU 进行比较，收到最优 BPDU 的端口 Port1 将成为根端口。

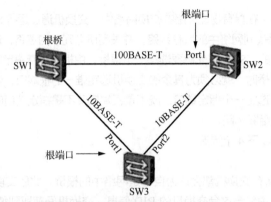

图 2-7　在非根桥上选举一个根端口

所谓根端口，简单理解就是非根桥上距离根桥最近的端口。根端口的选举分为以下 3 步。
（1）比较端口入 CoP 的大小，越小的越优先。
（2）比较端口对端设备 BID 的大小，越小的越优先。
（3）比较端口对端端口 PID 的大小，越小的越优先。

3. 指定端口选举

STP 将在每个网段中选举一个指定端口（Designated Port，DP），这个端口是该网段内所有端口中到达根桥的最优端口。此外，指定端口还负责向该网段发送 BPDU。对于非根桥而言，其所有端口中收到最优 BPDU 的端口将成为该非根桥的根端口，随后该非根桥使用自己接收的最优 BPDU，为本设备上的其他端口各计算一个 BPDU，然后使用计算出的 BPDU 与端口上所维护的 BPDU（端口自身也会从网络中收到 BPDU，并将 BPDU 保存起来）进行比较，如果前者更优，那么该端口将成为指定端口，且其所保存的 BPDU 也被前者替代，交换机将替代后的 BPDU 从该指定端口转发给下游交换机；如果后者更优，那么该端口将成为非指定端口（非指定端口既不是根端口，又不是指定端口）。

综上所述，对于非根桥而言，根端口的选举过程是非根桥对自己所收到的所有 BPDU 进行比较，而指定端口的选举过程则是非根桥用自己计算出的 BPDU 与其他设备发过来的 BPDU 进行比较。

如图 2-8 所示，在 SW1 与 SW2 之间的网段中，SW1 的 Port1 被选举为指定端口；在 SW1 与 SW3 之间的网段中，SW1 的 Port2 被选举为指定端口。一般而言，根桥的所有端口都是指定端口。另外，STP 还会在 SW2 与 SW3 之间的网段中选举一个指定端口，最终 SW2 的 Port2 端口胜出，成为该网段的指定端口。

所谓指定端口，简单理解就是每条链路上距离根桥最近的端口。指定端口的选举分为以下 3 步。
（1）比较端口处 CoP 的大小，越小的越优先。
（2）比较本端 BID 的大小，越小的越优先。
（3）比较本端 PID 的大小，越小的越优先。

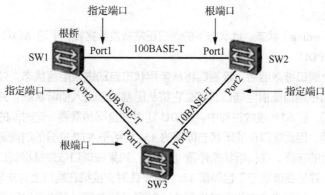

图 2-8 选举一个根指定端口

4. 阻塞非指定端口

经 STP 计算后，如果交换机的某个（或者某些）端口既不是根端口又不是指定端口，则将这种选举剩余下来的端口称为阻塞非指定端口（Non-Designated Port，NDP），那么该端口将会被 STP 阻塞，如此网络中的二层环路也就被打破了，至此 STP 收敛结束。

如图 2-9 所示，SW3 的 Port2 由于既不是根端口，又不是指定端口而被阻塞。被阻塞的端口既不会接收又不会转发业务数据（业务数据有别于 BPDU，可以简单地理解为网络中 PC 等设备发送的应用数据），且该端口不会发送 BPDU，但是会持续侦听 BPDU，以便感知网络拓扑的变更情况。

总之，每条链路一定有一个指定端口，根端口对面一定是指定端口，根桥上没有根端口，根桥上的所有端口都是指定端口。

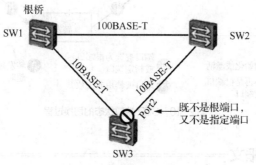

图 2-9 阻塞非指定端口

2.1.4 STP 的端口状态

STP 的端口状态主要有以下几种。

（1）禁用（Disabled）状态。此状态下的端口不能收发数据帧，不能收发 BPDU，不学习 MAC 地址表，不参与 STP 计算。

（2）阻塞（Blocking）状态。此状态下的端口不转发数据帧，不学习 MAC 地址表，接收并处理 BPDU，但是不向外发送 BPDU。

（3）侦听（Listening）状态。此状态下的端口不转发数据帧，不学习 MAC 地址表，只参与 STP 计算，接收并发送 BPDU。

（4）学习（Learning）状态。此状态下的端口不转发数据帧，但是学习 MAC 地址表，参与 STP 计算，接收并发送 BPDU。

（5）转发（Forwarding）状态。此状态下的端口正常转发数据帧，学习 MAC 地址表，参与 STP 计算，接收并发送 BPDU。

当交换机的一个端口被激活后，该端口将从禁用状态自动进入阻塞状态，处于阻塞状态的端口如果被交换机选举为根端口或指定端口，那么它将从阻塞状态进入侦听状态，并且在侦听状态停留 15s（转发延迟时间）。在 STP 收敛过程中，BPDU 泛洪到全网是需要一定时间的，STP 完成全网拓扑计算同样需要时间，因此端口在侦听状态停留的 15s 将给予 STP 充分的时间进行全网拓扑计算，避免网络中出现临时的环路。在侦听状态停留 15s 后，如果该端口依然是根端口或指定端口，那么它将进入学习状态，并且在该状态下也停留 15s。由于此时交换机在端口上并未学习到任何 MAC 地址，因此如果端口从侦听状态立即进入转发状态，就有可能在短时间内导致网络中出现不必要的数据帧泛洪现象。所以，STP 规定端口从侦听状态进入学习状态后需停留 15s，在这段时间内，交换机会在该端口持续侦听数据帧并学习 MAC 地址，为进入转发状态做好准备。图 2-10 描述了 STP 端口状态的切换过程。

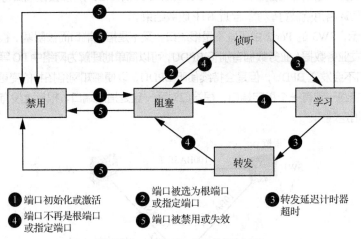

图 2-10　STP 端口状态的切换过程

2.1.5　STP 的报文

在一个交换网络中，STP 能够正常工作的基本前提是 BPDU 的正常交互。STP 的 BPDU 有两种类型：配置 BPDU（Configuration BPDU）和拓扑变化通知 BPDU（Topology Change Notification BPDU，TCN BPDU）。

1. Configuration BPDU 报文

Configuration BPDU 包含 BID、CoP 和 PID 等参数。STP 通过在交换机之间传递 Configuration BPDU 来选举根桥，以及确定每个交换机端口的角色和状态。在初始形成 STP 的过程中，各 STP 交换机都会周期性地主动产生和发送 Configuration BPDU。在 STP 形成后，只有根桥才会周期性地主动产生和发送 Configuration BPDU；相应地，非根桥会从自己的根端口周期性地接收 Configuration BPDU，并被立即触发而产生自己的 Configuration BPDU，同时将其从自己的指定端口发送出去。Configuration BPDU 报文格式如表 2-1 所示。

表 2-1 Configuration BPDU 报文格式

字段名	长度/字节	说明
Protocol Identifier（协议 ID）	2	总是为 0
Protocol Version（协议版本）	1	表示 STP 时，其值为 0；表示 RSTP 时，其值为 2；表示 MSTP 时，其值为 3
Message Type（消息类型）	1	指示当前 BPDU 消息类型：0x00 为 Configuration BPDU，0x80 为 TCN BPDU
Flags（标志）	1	最低位为拓扑变化（Topology Change，TC）标志，最高位为拓扑变化确认（Topology Change Acknowledgement，TCA）标志
Root Identifier（根 ID）	8	指示当前根桥的 BID（即"根 ID"），由 2 字节的桥优先级和 6 字节的 MAC 地址构成
Root Path Cost（根路径开销）	4	指示发送该 BPDU 报文的端口累计到根桥的开销
Bridge Identifier（根 ID）	8	指示发送该 BPDU 报文的交换设备的 BID（即"发送者 BID"），也是由 2 字节的桥优先级和 6 字节的 MAC 地址构成的
Port Identifier（端口 ID）	2	指示发送该 BPDU 报文的端口 ID，即"发送端口 ID"
Message Age（消息生存时间）	2	指示该 BPDU 报文的生存时间，即端口保存 BPDU 的最长时间，过期后将删除，要在这个时间内转发才有效。如果 Configuration BPDU 是直接来自根桥的，则 Message Age 为 0；如果是其他交换机转发的，则 Message Age 是指从根桥发送到当前交换机接收到 BPDU 的总时间，包括传输延时等。实际实现中，Configuration BPDU 报文经过一台交换机后，Message Age 会增加 1
Max Age（最大生存时间）	2	指示 BPDU 的最大生存时间，即老化时间
Hello Time（Hello 消息计时器）	2	指示发送两个相邻 BPDU 的时间间隔
Forward Delay（转发延时）	2	指示控制侦听和学习状态的持续时间，表示在拓扑结构改变后，交换机在发送数据包前维持在侦听和学习状态的时间

在 Configuration BPDU 报文中，Message Type 的值被设置为 0x00，主要作用如下。

（1）用于选举根桥及端口角色。

（2）通过定期发送（每 2s 发送一次）Configuration BPDU 报文维护端口状态。

（3）用于确认接收的 TCN BPDU 报文。

Configuration BPDU 中携带的参数类型有 3 种：对自身的标识参数、用于进行 STP 计算的参数和时间参数。

① 对自身的标识参数：包括 Protocol Identifier、Protocol Version、Message Type 和 Flags。

② 用于进行 STP 计算的参数：包括发送该 BPDU 的交换机的 BID、当前根桥的 BID、发送该 BPDU 的端口的 PID、发送该 BPDU 的端口的根路径开销（Root Path Cost，RPC）。

③ 时间参数：包括 Hello Time、Forward Delay、Message Age、Max Age。

2. TCN BPDU 报文

TCN BPDU 的格式非常简单，只有表 2-1 中的 Protocol Identifier、Protocol Version、Message Type 这 3 个字段，并且 Message Type 字段的值为 0x80。TCN BPDU 用于在网络拓扑发生变化时向根桥通知变化的发生。

STP 在以下 3 种情况下会发送 TCN BPDU 报文。

（1）端口从转发状态过渡到阻塞状态或者禁用状态。

（2）非根桥从一个指定端口收到 TCN BPDU 报文后会从自己的根端口向根桥转发。

（3）端口进入转发状态并且交换机已经存在一个指定端口。

TCN BPDU 是用于维护 STP 的稳定的，当网络拓扑发生变化时，直连的交换机可以直接感受到故障点，于是会发送 TCN BPDU 给上行交换机，上行交换机收到后会回应一个 TCA 标志置 1 的 Configuration BPDU。上行交换机会以发送 Hello Time 为周期通过根端口不断向它的上游交换机发送 TCN BPDU。这个过程不断重复，直到根桥接收到 TCN BPDU。根桥接收到 TCN BPDU 后，会发送 TC 标志置 1 的 Configuration BPDU 来通告所有交换机整个网络拓扑已经发生了变化。非根桥收到 TC 标志置 1 的 Configuration BPDU 后，就说明自身的 MAC 地址表项可能不是正确的，于是将自己的 MAC 地址表的 Max Age 缩短为 Forward Delay 的时间长度，以加速老化原来的 MAC 地址表项，如图 2-11 所示。

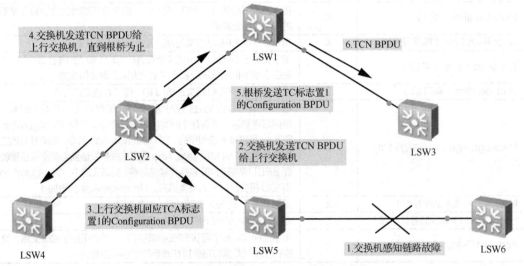

图 2-11　根端口发送 TCN BPDU 给上行交换机

2.1.6　STP 的计时器

在 STP 的计算过程中，用到了 BPDU 报文的 Hello 消息计时器（Hello Time）、转发延时（Forward Delay）和最大生存时间（Max Age）这 3 个主要时间参数。

1. Hello Time

Hello Time 是运行 STP 的设备发送 Configuration BPDU 的时间间隔，默认为 2s，用于设备检测链路是否存在故障。当网络拓扑稳定之后，该计时器的修改只有在根桥修改后才有效。新的根桥会在发出的 BPDU 报文中填充相应的字段以向其他非根桥传递该计时器修改的信息。但当网络拓扑变化之后，TCN BPDU 的发送不受这个计时器的管理。

2. Forward Delay

Forward Delay 是指设备状态迁移的延时。链路故障会引发网络重新进行 STP 的计算，STP 的结构将发生相应的变化。不过重新计算得到的新 Configuration BPDU 无法立刻传遍整个网络，如果新选出的根端口和指定端口立刻开始数据转发，则可能会造成临时环路。为此，STP 采用了一种状态迁移机制，新选出的根端口和指定端口要经过 2 倍的 Forward Delay 后才能进入转发状态，这个延时保证了新的 Configuration BPDU 传遍整个网络，从而防止了临时环路的产生。

Forward Delay 指一个端口处于侦听和学习状态的各自持续时间，默认是 15s。即侦听状态持续

15s,随后学习状态再持续 15s。这两个状态下的端口会处于阻塞状态,这正是 STP 用于避免临时环路的关键。

3. Max Age

Max Age 是指端口 BPDU 报文的最大生存时间,默认为 20s,可在根桥上通过命令手动修改 Max Age。

Max Age 通过 Configuration BPDU 报文的传输,可保证 Max Age 在整网中一致。运行 STP 的网络中的非根桥收到 Configuration BPDU 报文后,会对其中的 Message Age 和 Max Age 进行比较:如果 Message Age≤Max Age,则该非根桥继续转发 Configuration BPDU 报文;如果 Message Age>Max Age,则该 Configuration BPDU 报文将被老化,该非根桥直接丢弃该 Configuration BPDU,可认为网络直径过大,导致根桥连接失败。

如图 2-12 所示,S2 和 S3 从 S1 收到 Configuration BPDU,Message Age 为 0,所以在 S2 和 S3 去往 S1 的端口上,Configuration BPDU 报文的最大生存时间为(Max Age-0)。

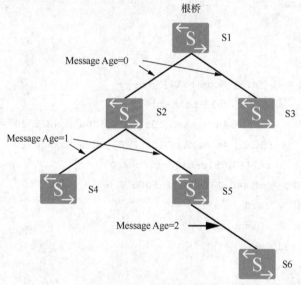

图 2-12 Message Age 示意

S4 和 S5 从 S2 收到 Configuration BPDU,Message Age 为 1,所以在 S4 和 S5 去往 S1 的端口上,Configuration BPDU 报文的最大生存时间为(Max Age-1)。

S6 从 S5 收到 Configuration BPDU,Message Age 为 2,所以在 S6 去往 S1 的端口上,Configuration BPDU 报文的最大生存时间为(Max Age-2)。

修改计时器的过程如下。

(1)执行 system-view 命令,进入系统视图。

(2)执行 stp timer forward-delay forward-delay 命令,配置设备的 Forward Delay。默认情况下,设备的 Forward Delay 是 1500cs[15s,1s=100cs(厘秒)]。

(3)执行 stp timer hello hello-time 命令,配置设备的 Hello Time。默认情况下,设备的 Hello Time 是 200cs(2s)。

(4)执行 stp timer max-age max-age 命令,配置设备的 Max Age。默认情况下,设备的 Max Age 是 2000cs(20s)。

具体代码如下。

```
<HUAWEI> system-view
[HUAWEI] stp timer forward-delay 2000    //配置Forward Delay为2000cs
[HUAWEI] stp timer hello 400             //配置Hello Time为400cs
[HUAWEI] stp timer max-age 1000          //配置Max Age为1000cs
[HUAWEI] display stp global
Protocol Status            : Enabled
Bpdu-filter default        : Disabled
Tc-protection              : Enabled
Tc-protection threshold    : 1
Tc-protection interval     : 2s
Edged port default         : Enabled
Pathcost-standard          : Dot1t
Timer-factor               : 3
Transmit-limit             : 10
Bridge-diameter            : 7
-------[CIST Global Info][Mode MSTP]-------
CIST Bridge            :61440.781d-ba56-f06c
Config Times           :Hello 4s MaxAge 10s FwDly 20s MaxHop 20
Active Times           :Hello 4s MaxAge 10s FwDly 20s MaxHop 20
CIST Root/ERPC         :61440.781d-ba56-f06c / 0
CIST RegRoot/IRPC      :61440.781d-ba56-f06c / 0
CIST RootPortId        :0.0
BPDU-Protection        :Disabled
TC or TCN received     :85
TC count per hello     :0
STP Converge Mode      :Normal
Share region-configuration :Enabled
Time since last TC     :0 days 9h:12m:34s
Number of TC           :13
Last TC occurred       :GigabitEthernet0/0/4
```

【任务实施】

实验1 配置STP

实验目的：
（1）熟悉STP的工作原理。
（2）理解根桥、根端口、指定端口的作用。
（3）掌握根桥、根端口、指定端口的选举条件。

V2-1 配置 STP-1

V2-2 配置 STP-2

STP 有 STP、RSTP、MSTP 这 3 种类型。华为设备默认运行的是 MSTP。如果要配置 STP，则所有交换机都必须调整为 STP 版本。本实验模拟企业网络场景，由 4 台交换机两两相连组成一个环形网络，为了避免形成环路，每台交换机都运行了 STP，且配置了 SW1 为根桥，SW3 为备份桥。现在为了优化网络，在网络变化时加快 STP 的收敛速度，需要在交换机上更改 STP 计时器的设置，将所有计时器调整到最优值，完成 STP 的加速收敛。图 2-13 所示为 STP 详细参数示意，其中，"ROOT"表示根桥、"DP"表示指定端口、"RP"表示"根端口"，"ATL"表示阻塞端口。

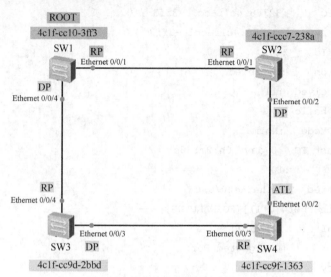

图 2-13 STP 详细参数示意

实验步骤如下。
（1）所有交换机调整为 STP 版本，通过使用 stp mode stp 命令进行设置，代码如下。

```
<Huawei>sys
Enter system view, return user view with Ctrl+Z.
[Huawei]sys SW1
[SW1]
Sep 11 2021 11:39:49-08:00 SW1 DS/4/DATASYNC_CFGCHANGE:OID 1.3.6.1.4.1.2011.5.25.
191.3.1 configurations have been changed. The current change number is 4, the c
hange loop count is 0, and the maximum number of records is 4095.
[SW1]stp mode ?
  mstp  Multiple Spanning Tree Protocol (MSTP) mode
  rstp  Rapid Spanning Tree Protocol (RSTP) mode
  stp   Spanning Tree Protocol (STP) mode
[SW1]stp mode stp
Info: This operation may take a few seconds. Please wait for a moment...done.

[SW2]stp mode stp
[SW3]stp mode stp
[SW4]stp mode stp
```

（2）查看 STP 的完整信息，代码如下。

```
[SW1]display stp
-------[CIST Global Info][Mode STP]-------
CIST Bridge            :32768.4c1f-cc10-3ff3
Config Times           :Hello 2s MaxAge 20s FwDly 15s MaxHop 20
Active Times           :Hello 2s MaxAge 20s FwDly 15s MaxHop 20
CIST Root/ERPC         :32768.4c1f-cc10-3ff3 / 0
CIST RegRoot/IRPC      :32768.4c1f-cc10-3ff3 / 0
CIST RootPortId        :0.0
BPDU-Protection        :Disabled
TC or TCN received     :11
TC count per hello     :0
STP Converge Mode      :Normal
Time since last TC     :0 days 0h:2m:36s
Number of TC           :10
Last TC occurred       :Ethernet0/0/4
----[Port1(Ethernet0/0/1)][FORWARDING]----
 Port Protocol         :Enabled
 Port Role             :Designated Port
 Port Priority         :128
 Port Cost(Dot1T )     :Config=auto / Active=200000
 Designated Bridge/Port    :32768.4c1f-cc10-3ff3 / 128.1
 Port Edged            :Config=default / Active=disabled
 Point-to-point        :Config=auto / Active=true
 Transit Limit         :147 packets/hello-time
 Protection Type       :None
---- More ----
```

（3）寻找根桥。通过使用 display interface Vlanif 1 命令查看所有交换机的硬件地址，地址最小的即为根桥。如图 2-13 所示，SW1 为根桥，代码如下。

```
[SW1]display interface Vlanif 1
Vlanif1 current state : UP
Line protocol current state : DOWN
Description:
Route Port, The Maximum Transmit Unit is 1500
Internet protocol processing : disabled
IP Sending Frames' Format is PKTFMT_ETHNT_2, Hardware address is 4c1f-cc10-3ff3
Current system time: 2021-09-11 11:46:19-08:00
    Input bandwidth utilization  : --
Output bandwidth utilization : --


[SW2]display interface Vlanif 1
```

```
Vlanif1 current state : UP
Line protocol current state : DOWN
Description:
Route Port, The Maximum Transmit Unit is 1500
Internet protocol processing : disabled
IP Sending Frames' Format is PKTFMT_ETHNT_2, Hardware address is 4c1f-ccc7-238a
Current system time: 2021-09-11 11:48:45-08:00
    Input bandwidth utilization  : --
Output bandwidth utilization : --

[SW3]display interface Vlanif 1
Vlanif1 current state : UP
Line protocol current state : DOWN
Description:
Route Port, The Maximum Transmit Unit is 1500
Internet protocol processing : disabled
IP Sending Frames' Format is PKTFMT_ETHNT_2, Hardware address is 4c1f-cc9d-2bbd
Current system time: 2021-09-11 11:49:25-08:00
    Input bandwidth utilization  : --
    Output bandwidth utilization : --

[SW4]display interface Vlanif 1
Vlanif1 current state : UP
Line protocol current state : DOWN
Description:
Route Port, The Maximum Transmit Unit is 1500
Internet protocol processing : disabled
IP Sending Frames' Format is PKTFMT_ETHNT_2, Hardware address is 4c1f-cc9f-1363
Current system time: 2021-09-11 11:49:44-08:00
    Input bandwidth utilization  : --
Output bandwidth utilization : --
```

（4）查看 STP 的简要信息。通过使用 display stp brief 命令查看所有交换机的 STP 简要信息，找出根端口（ROOT）、指定端口（DEST）和阻塞端口（ALTE），代码如下。

```
[SW1]display stp brief
 MSTID   Port                  Role  STP State   Protection
   0     Ethernet0/0/1         DESI  FORWARDING  NONE
   0     Ethernet0/0/4         DESI  FORWARDING  NONE


[SW2]display stp brief
 MSTID   Port                  Role  STP State   Protection
   0     Ethernet0/0/1         ROOT  FORWARDING  NONE
```

```
    0           Ethernet0/0/2           DESI    FORWARDING      NONE


[SW3]display stp brief
 MSTID  Port                            Role    STP State       Protection
    0           Ethernet0/0/3           DESI    FORWARDING      NONE
    0           Ethernet0/0/4           ROOT    FORWARDING      NONE


[SW4]display stp brief
 MSTID  Port                            Role    STP State       Protection
    0           Ethernet0/0/2           ALTE    DISCARDING      NONE
    0           Ethernet0/0/3           ROOT    FORWARDING      NONE
```

图 2-13 中交换机 SW1 的 Ethernet 0/0/1、SW2 的 Ethernet 0/0/1、交换机 SW3 的 Ethernet 0/0/4、交换机 SW4 的 Ethernet 0/0/3 为根端口，交换机 SW1 的 Ethernet 0/0/4、交换机 SW2 的 Ethernet 0/0/2、交换机 SW3 的 Ethernet 0/0/3 为指定端口，交换机 SW4 的 Ethernet 0/0/2 为阻塞端口。由实验可知，根桥上所有端口都为指定端口，每台非根桥上只有一个唯一的根端口，每个根端口的对端端口一定为指定端口。

假设将 SW2 设置为新的根桥，STP 角色更改的拓扑示意如图 2-14 所示。每台交换机的优先级都为 32768，优先级的值越小越优先，因此可以把 SW2 的优先级降低，修改为小于 32768 的数值，此时 SW2 就变为根桥。

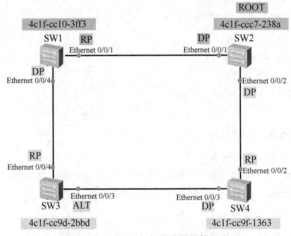

图 2-14　STP 角色更改的拓扑示意

（5）查看优先级。通过使用 display stp 命令可以查看到所有设备的优先级都为 32768，代码如下。

```
[SW1]display stp
-------[CIST Global Info][Mode STP]-------
CIST Bridge            :32768.4c1f-cc10-3ff3
Config Times           :Hello 2s MaxAge 20s FwDly 15s MaxHop 20
Active Times           :Hello 2s MaxAge 20s FwDly 15s MaxHop 20
CIST Root/ERPC         :32768.4c1f-cc10-3ff3 / 0
CIST RegRoot/IRPC      :32768.4c1f-cc10-3ff3 / 0
CIST RootPortId        :0.0
```

```
BPDU-Protection     :Disabled
TC or TCN received  :6
TC count per hello  :0
STP Converge Mode   :Normal
Time since last TC  :0 days 0h:0m:58s
Number of TC        :4
Last TC occurred    :Ethernet0/0/4
……


[SW2]display stp
-------[CIST Global Info][Mode STP]-------
CIST Bridge         :32768.4c1f-ccc7-238a
Config Times        :Hello 2s MaxAge 20s FwDly 15s MaxHop 20
Active Times        :Hello 2s MaxAge 20s FwDly 15s MaxHop 20
CIST Root/ERPC      :32768.4c1f-cc10-3ff3 / 200000
CIST RegRoot/IRPC   :32768.4c1f-ccc7-238a / 0
CIST RootPortId     :128.1
BPDU-Protection     :Disabled
TC or TCN received  :6
TC count per hello  :0
STP Converge Mode   :Normal
Time since last TC  :0 days 0h:1m:54s
Number of TC        :4
Last TC occurred    :Ethernet0/0/1
……

[SW3]display stp
-------[CIST Global Info][Mode STP]-------
CIST Bridge         :32768.4c1f-cc9d-2bbd
Config Times        :Hello 2s MaxAge 20s FwDly 15s MaxHop 20
Active Times        :Hello 2s MaxAge 20s FwDly 15s MaxHop 20
CIST Root/ERPC      :32768.4c1f-cc10-3ff3 / 200000
CIST RegRoot/IRPC   :32768.4c1f-cc9d-2bbd / 0
CIST RootPortId     :128.4
BPDU-Protection     :Disabled
TC or TCN received  :4
TC count per hello  :0
STP Converge Mode   :Normal
Time since last TC  :0 days 0h:2m:32s
Number of TC        :4
```

```
Last TC occurred       :Ethernet0/0/4
……
[SW4]display stp
-------[CIST Global Info][Mode STP]-------
CIST Bridge            :32768.4c1f-cc9f-1363
Config Times           :Hello 2s MaxAge 20s FwDly 15s MaxHop 20
Active Times           :Hello 2s MaxAge 20s FwDly 15s MaxHop 20
CIST Root/ERPC         :32768.4c1f-cc10-3ff3 / 400000
CIST RegRoot/IRPC      :32768.4c1f-cc9f-1363 / 0
CIST RootPortId        :128.3
BPDU-Protection        :Disabled
TC or TCN received     :7
TC count per hello     :0
STP Converge Mode      :Normal
Time since last TC     :0 days 0h:3m:0s
Number of TC           :4
Last TC occurred       :Ethernet0/0/3
……
```

（6）设置 SW2 为新的根桥。修改 SW2 的优先级为 $4096 \times n$（n 为整数），且小于 32768，那么 SW2 将成为新的根桥，代码如下。

```
[SW2]stp priority 4096
[SW2]display stp
Sep 14 2021 10:49:02-08:00 SW2 DS/4/DATASYNC_CFGCHANGE:OID 1.3.6.1.4.1.2011.5.25
.191.3.1 configurations have been changed. The current change number is 5, the c
hange loop count is 0, and the maximum number of records is 4095.
-------[CIST Global Info][Mode MSTP]-------
CIST Bridge            :4096.4c1f-ccc7-238a
Config Times           :Hello 2s MaxAge 20s FwDly 15s MaxHop 20
Active Times           :Hello 2s MaxAge 20s FwDly 15s MaxHop 20
CIST Root/ERPC         :4096.4c1f-ccc7-238a / 0
CIST RegRoot/IRPC      :4096.4c1f-ccc7-238a / 0
CIST RootPortId        :0.0
BPDU-Protection        :Disabled
TC or TCN received     :10
TC count per hello     :0
STP Converge Mode      :Normal
Time since last TC     :0 days 0h:0m:4s
Number of TC           :6
Last TC occurred       :Ethernet0/0/1
```

（7）再次查看 STP 的简要信息。各端口的状态信息如下。

```
[SW1]display stp brief
 MSTID  Port              Role  STP State   Protection
   0    Ethernet0/0/1     ROOT  FORWARDING  NONE
   0    Ethernet0/0/4     DESI  FORWARDING  NONE

[SW2]display stp brief
 MSTID  Port              Role  STP State   Protection
   0    Ethernet0/0/1     DESI  FORWARDING  NONE
   0    Ethernet0/0/2     DESI  FORWARDING  NONE

[SW3]display stp brief
 MSTID  Port              Role  STP State   Protection
   0    Ethernet0/0/3     ALTE  DISCARDING  NONE
   0    Ethernet0/0/4     ROOT  FORWARDING  NONE

[SW4]display stp brief
 MSTID  Port              Role  STP State   Protection
   0    Ethernet0/0/2     ROOT  FORWARDING  NONE
   0    Ethernet0/0/3     DESI  FORWARDING  NONE
```

（8）修改 SW3 的 Ethernet 0/0/3 端口为新的根端口。假如想使 SW3 的两个端口角色互换，也就是把 Ethernet 0/0/3（原来为阻塞端口）修改为根端口（根端口可以比较端口对端设备的 BID 值的大小，BID 的值越小越优先），则可以修改 Ethernet 0/0/3 端口对端的交换机 SW4 的优先级为 8192，如图 2-15 所示，代码如下。

```
[SW4]stp priority 8192
[SW4]display stp
-------[CIST Global Info][Mode MSTP]-------
CIST Bridge         :8192.4c1f-cc9f-1363
Config Times        :Hello 2s MaxAge 20s FwDly 15s MaxHop 20
Active Times        :Hello 2s MaxAge 20s FwDly 15s MaxHop 20
CIST Root/ERPC      :4096.4c1f-ccc7-238a / 200000
CIST RegRoot/IRPC   :8192.4c1f-cc9f-1363 / 0
CIST RootPortId     :128.2
BPDU-Protection     :Disabled
TC or TCN received  :10
TC count per hello  :0
STP Converge Mode   :Normal
Time since last TC  :0 days 0h:0m:6s
Number of TC        :7
Last TC occurred    :Ethernet0/0/3
```

```
……
[SW3]display stp brief
MSTID    Port                    Role    STP State      Protection
0        Ethernet0/0/3           ROOT    LEARNING       NONE
0        Ethernet0/0/4           ALTE    DISCARDING     NONE
```

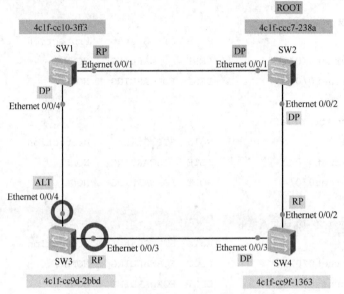

图 2-15　SW3 两个端口角色互换拓扑示意

【任务拓展】

（1）图 2-16 所示为 STP 拓展实验拓扑示意，请给图 2-16 中的各个设备标注默认角色（交换机、根端口、指定端口、非指定端口）。

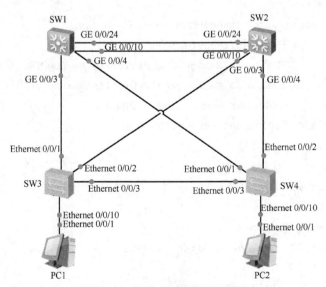

图 2-16　STP 拓展实验拓扑示意

（2）修改 STP 部分参数（如更改根桥、根端口等）后，检查各交换机和端口角色的变化，重新在图 2-16 中给各设备标注修改后的角色（交换机、根端口、指定端口、非指定端口）。

任务 2.2　VLAN 划分

【知识准备】

2.2.1　VLAN 相关介绍

1. 冲突域

冲突域是指连接在同一线缆上的所有工作站的集合，或者说是同一物理网段上所有节点的集合，又或者说是以太网上竞争同一带宽的节点集合。这个域代表了冲突在其中发生并传播的区域，这个区域可以被认为是共享段。在 OSI 参考模型中，冲突域被看作第一层的概念，连接同一冲突域的设备有集线器、中继器或者其他进行简单复制信号的设备。用集线器或者中继器连接的所有节点可以被认为是在同一个冲突域内。冲突域如图 2-17 所示。

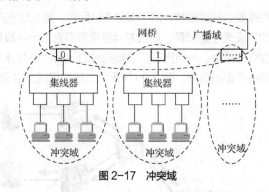

图 2-17　冲突域

2. 广播域

广播域是指接收同样广播消息的节点的集合。如在该集合中的任何一个节点传输一个广播帧，则所有其他能收到这个帧的节点都被认为是该广播域的一部分。许多设备都极易产生广播，如果不维护，就会消耗大量的带宽，降低网络工作效率。因为广播域被认为是 OSI 参考模型中的第二层概念，所以以集线器、网桥等第一层、第二层设备连接的节点被认为在同一个广播域。而路由器、三层交换机则可以划分广播域，即可以连接不同的广播域。广播域如图 2-18 所示。

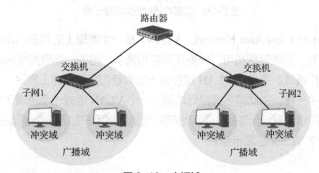

图 2-18　广播域

3. VLAN 的概念

对于一台交换机而言，默认情况下它的所有端口都属于同一个广播域，当多台主机连接到同一台交换机时，它们可以直接进行通信（只需配置相同网段的 IP 地址），而无须借助路由设备，这种通信行为被称为二层通信。由于这些主机都属于同一个广播域，因此当其中一台主机发出一个广播帧时，连接在交换机上的其他所有主机都会收到这个帧。当然，如果交换机在某个端口上收到目的 MAC 地址未知的单播帧，则会对这个帧进行泛洪广播。所有主机属于同一广播域如图 2-19 所示。

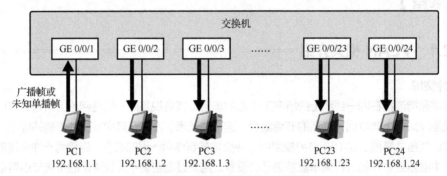

图 2-19　所有主机属于同一广播域

然而，并非所有主机都需要这些数据帧，对于部分主机而言，接收这些广播帧或者目的 MAC 地址未知的单播帧实际上增加了设备性能损耗，而且对网络带宽而言是一种浪费。例如，在一个由许多二层交换机构成的大型二层广播域中，一旦出现广播帧或目的 MAC 地址未知的单播帧就会引发大量的泛洪现象，从而给网络带来沉重的负担。图 2-20 所示为二层广播域泛洪现象示意。

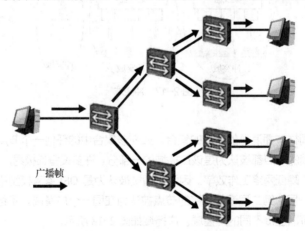

图 2-20　二层广播域泛洪现象示意

虚拟局域网（Virtual Local Area Network，VLAN）是一组逻辑上的设备和用户，这些设备和用户并不受物理位置的限制，这使管理员可以根据实际应用需求，把同一物理局域网内的不同设备和用户逻辑地划分成不同的广播域。每一个 VLAN 都包含一组有着相同需求的工作站，它们之间的通信就好像它们在同一个网段中一样，由此得名虚拟局域网。一个 VLAN 组成一个逻辑子网，即一个逻辑广播域，它可以覆盖多个网络设备，允许处于不同地理位置的网络用户加入一个逻辑子网中。图 2-21 所示为不同楼层划分同一个 VLAN 区域示意。

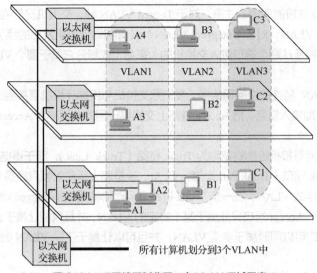

图 2-21　不同楼层划分同一个 VLAN 区域示意

4. VLAN 的优点

VLAN 的优点如下。

（1）防范广播风暴：一个 VLAN 就是一个逻辑广播域，通过对 VLAN 的创建，隔离了广播，缩小了广播范围，可以控制广播风暴的产生。将网络划分为多个 VLAN 可减少参与广播风暴的设备数量，VLAN 分段可以防止广播风暴波及整个网络。VLAN 可以提供防火墙建立的机制，防止交换网络的过量广播。使用 VLAN 可以将交换端口或计算机指定到某一个特定的 VLAN，该 VLAN 可以在一台交换机中，也可跨多台交换机，一个 VLAN 中的广播不会发送到 VLAN 之外，这样可以把广播控制在一个 VLAN 内，减少广播造成的影响。

（2）提高网络安全性：将不同用户群划分在不同 VLAN，可增强局域网的安全性，含有敏感数据的用户组可与网络的其余部分隔离，从而降低泄露机密信息的可能性。不同 VLAN 内的报文在传输时是相互隔离的，即一个 VLAN 内的用户不能和其他 VLAN 内的用户直接通信。如果不同 VLAN 内的用户要进行通信，则需要使用路由器或三层交换机等第三层设备。

（3）降低成本：成本高昂的网络升级需求减少，现有带宽和上行链路的利用率更高，从而可节约成本。

（4）提高性能：将第二层平面网络划分为多个逻辑工作组（广播域）可以减少网络上不必要的流量并提高性能。

（5）提高人员工作效率：VLAN 为网络管理带来了方便，因为有相似网络需求的用户将共享同一个 VLAN。

2.2.2　链路类型和端口类型

1. 链路类型

根据链路中需要承载的 VLAN 数目的不同，以太网链路分为接入链路和干道链路。

一个 VLAN 帧可能带有 Tag（称为 Tagged VLAN 帧，或称为 Tagged 帧），也可能不带 Tag（称为 Untagged VLAN 帧，或称为 Untagged 帧）。在谈及 VLAN 技术时，如果一个帧被交换机划分到 VLANi（i=1,2,3,…,4094），则就把这个帧称为一个 VLANi 帧。对于带有 Tag 的 VLANi 帧，i 其实就是这个帧的 Tag 中的 VLAN ID 字段的取值。注意，对于 Tagged VLAN 帧，交换机显然能够从其 Tag 中的 VID 值判定出它属于哪个 VLAN；对于 Untagged VLAN 帧（如终端计算机发出的帧），交换机需要根据某种原则（如根据这个帧是从哪个端口进入交换机的）来判定或划分它属于哪个 VLAN。

（1）接入链路

在一个支持 VLAN 特性的交换网络中，我们把交换机与终端计算机直接相连的链路称为 Access 链路（Access Link），即接入链路，把 Access 链路上交换机一侧的端口称为 Access 端口（Access Port）。

（2）干道链路

我们把交换机之间直接相连的链路称为 Trunk 链路（Trunk Link），即干道链路，把 Trunk 链路上两侧的端口称为 Trunk 端口（Trunk Port）。在一条 Access 链路上运动的帧只能是 Untagged 帧，并且这些帧只能属于某个特定的 VLAN；在一条 Trunk 链路上运动的帧只能是 Tagged 帧，并且这些帧可以属于不同的 VLAN。一个 Access 端口只能属于某个特定的 VLAN，并且只能让属于这个特定 VLAN 的帧通过；一个 Trunk 端口可以同时属于多个 VLAN，并且可以让属于不同 VLAN 的帧通过。

2. 端口类型

每一台交换机的端口（无论是 Access 端口还是 Trunk 端口）都应该配置一个 PVID（Port VLAN ID），到达这个端口的 Untagged 帧将一律被交换机划分到 PVID 所指代的 VLAN。例如，如果一个端口的 PVID 被配置为 10，则所有到达这个端口的 Untagged 帧都将被认定为是属于 VLAN 10 的帧。在默认情况下，PVID 的值为 1。

概括地讲，链路上运动的帧有可能是 Tagged 帧，也有可能是 Untagged 帧。但一台交换机内部不同端口之间运动的帧一定是 Tagged 帧。

（1）Access 端口

Access 端口一般用于和不能识别 Tag 的用户终端（如用户主机、服务器等）相连，或者不需要区分不同 VLAN 成员时使用。它只能收发 Untagged 帧且只能为 Untagged 帧添加唯一 VLAN 的 Tag。

（2）Trunk 端口

Trunk 端口一般用于连接交换机、路由器、接入点（Access Point，AP），以及可同时收发 Tagged 帧和 Untagged 帧的语音终端，它可以允许多个 VLAN 的帧带 Tag 通过，但只允许一个 VLAN 的帧从该类端口上发出时不带 Tag（即剥除 Tag）。

（3）Hybrid 端口

Hybrid 端口为混合端口，既可以用于连接不能识别 Tag 的用户终端（如用户主机、服务器等）和网络设备（如集线器、"傻瓜式"交换机），也可以用于连接路由器及可同时收发 Tagged 帧和 Untagged 帧的语音终端、AP。它可以允许多个 VLAN 的帧带 Tag 通过，并且允许从该类端口发出的帧根据需求配置某些 VLAN 的帧带 Tag（即不剥除 Tag）、某些 VLAN 的帧不带 Tag（即剥除 Tag）。Hybrid 端口和 Trunk 端口在很多应用场景下可以通用，但在某些应用场景下，必须使用 Hybrid 端口，如在一个端口连接不同 VLAN 网段的场景中，因为一个端口需要给多个 Untagged 帧添加 Tag，所以必须使用 Hybrid 端口。

2.2.3 划分 VLAN 的基本方法

从技术角度讲，VLAN 的划分可依据不同原则，一般有 3 种划分方法，分别是基于端口的 VLAN 划分、基于 MAC 地址的 VLAN 划分，以及基于路由的 VLAN 划分。

1. 基于端口的 VLAN 划分

这种划分方法是把一台或多台交换机上的几个端口划分为一个逻辑组，这是最简单、最有效的划分方法。该方法只需网络管理员对网络设备的交换端口进行重新分配即可，不用考虑该端口所连接的设备。

2. 基于 MAC 地址的 VLAN 划分

MAC 地址其实就是指网卡的标识符，每一块网卡的 MAC 地址都是唯一且固化在网卡上的。MAC 地址由 12 位十六进制数表示，前 8 位为厂商标识，后 4 位为网卡标识。网络管理员可按 MAC 地址把一些站点划分为一个逻辑子网。

3. 基于路由的 VLAN 划分

路由协议工作在网络层，相应的工作设备有路由器和路由交换机（即三层交换机）。该方法允许一个 VLAN 跨越多台交换机，或一个端口位于多个 VLAN 中。

就目前来说，对于 VLAN 的划分主要采取上述第 1、3 种方法，第 2 种方法为辅助性的方案。本书主要介绍第 1 种方法。

2.2.4 VLAN 注册协议概述

虚拟局域网注册协议（GARP VLAN Registration Protocol，GVRP）是通用属性注册协议（Generic Attribute Registration Protocol，GARP）的一种应用，用来维护交换机中的 VLAN 动态注册信息，并传播该信息到其他交换机中，避免手动地一个一个去配置静态 VLAN，进而提高工作效率。

GVRP 有以下 3 种注册模式。

（1）FIXED：不允许动态 VLAN 在端口上注册或者注销，且只发送静态 VLAN 的声明消息。

（2）FORBIDDEN：不允许动态 VLAN 在端口上进行注册，同时删除端口上除了 VLAN 1 以外的所有 VLAN。

（3）NORMAL：允许静态 VLAN 和动态 VLAN 注册，同时发送静态 VLAN 和动态 VLAN 的信息。默认情况下，GVRP 端口注册的模式为 NORMAL。

【任务实施】

实验 2　配置 VLAN

实验目的：

（1）理解 VLAN 的原理。

（2）熟悉交换机端口的 3 种类型。

（3）掌握 VLAN 的 3 种端口类型划分配置方法。

（4）理解 GVRP 的作用及配置。

本实验模拟企业网络场景，企业内网是一个大的局域网，二层交换机放置在一楼，在一楼办公的部门有人事部，在二楼办公的部门有销售部和研发部。因为交换机组成的网络是一个广播域，所以交换机连接的所有主机都能互相通信。该企业的策略是要求同一部门内的主机可以互相通信，销售部与研发部之间不能通信，人事部可以和销售部、研发部两个部门通信。

V2-3　VLAN（Access）配置实验

1. VLAN（Access）配置实验

图 2-22 所示为 VLAN（Access）配置实验的拓扑示意，把连接到交换机 SW1 上的计算机 PC1、PC2 分别划分到 VLAN 10、VLAN 20 中，详细步骤如下。

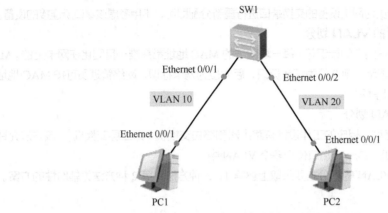

图 2-22 VLAN（Access）配置实验的拓扑示意

（1）创建 VLAN 10、VLAN 20，代码如下。

```
[SW1]VLAN 10
[SW1-VLAN 10]quit
[SW1]VLAN 20
[SW1-VLAN 20]quit
```

（2）查看设备的 VLAN 信息，代码如下。

```
[SW1]display vlan summary
static vlan:
Total 3 static vlan.
 1 10 20

dynamic vlan:
Total 0 dynamic vlan.

reserved vlan:
Total 0 reserved vlan.
```

（3）将端口划入对应的 VLAN，代码如下。

```
[SW1]int e0/0/1
[SW1-Ethernet0/0/1]port link-type Access
[SW1-Ethernet0/0/1]port default VLAN 10
[SW1-Ethernet0/0/1]quit
[SW1]int e0/0/2
[SW1-Ethernet0/0/2]port link-type Access
[SW1-Ethernet0/0/2]port default VLAN 20
[SW1-Ethernet0/0/2]quit
```

（4）查看端口对应的 VLAN 信息，代码如下。

```
[SW1]display port VLAN active
T=TAG U=UNTAG
```

```
--------------------------------------------------------------------
Port            Link Type     PVID     VLAN List
--------------------------------------------------------------------
Eth0/0/1        access        10       U: 10
Eth0/0/2        access        20       U: 20
Eth0/0/3        hybrid        1        U: 1
Eth0/0/4        hybrid        1        U: 1
Eth0/0/5        hybrid        1        U: 1
Eth0/0/6        hybrid        1        U: 1
……
```

2. VLAN（Trunk）配置实验

图 2-23 所示为 VLAN（Trunk）配置实验的拓扑示意，在两台交换机的主干线路两侧端口设置 Trunk 类型，建立 Trunk 链路，详细步骤如下。

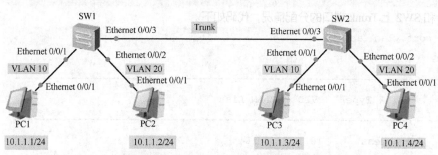

V2-4　VLAN（Trunk）配置实验

图 2-23　VLAN（Trunk）配置实验的拓扑示意

（1）在各交换机上创建 VLAN 10、VLAN 20，代码如下。

```
[SW1] vlan batch 10 20
[SW2] vlan batch 10 20
```

（2）将端口划入对应的 VLAN。如图 2-23 所示，把 SW1 的 Ethernet 0/0/1、Ethernet 0/0/2 分别加入 VLAN 10、VLAN 20，把 SW2 的 Ethernet 0/0/1、Ethernet 0/0/2 分别加入 VLAN 10、VLAN 20，代码如下。

```
[SW1]int Ethernet0/0/1
[SW1-Ethernet0/0/1]port link-type Access
[SW1-Ethernet0/0/1]port default VLAN 10
[SW1-Ethernet0/0/1]quit
[SW1]int e0/0/2
[SW1-Ethernet0/0/2]port default VLAN 20
[SW1-Ethernet0/0/2]quit

[SW2]int Ethernet0/0/1
[SW2-Ethernet0/0/1]port link-type Access
[SW2-Ethernet0/0/1]port default VLAN 10
[SW2-Ethernet0/0/1]quit
```

```
[SW2]int e0/0/2
[SW2-Ethernet0/0/2]port default VLAN 20
[SW2-Ethernet0/0/2]quit
```

（3）配置 Trunk 端口。分别把 SW1 和 SW2 的 Ethernet 0/0/3 设置为 Trunk 类型，代码如下。

```
[SW1]int Ethernet0/0/3
[SW1-Ethernet0/0/3]port link-type Trunk
[SW1-Ethernet0/0/3]port Trunk allow-pass VLAN all
[SW1-Ethernet0/0/3]quit


[SW2]int Ethernet0/0/3
[SW2-Ethernet0/0/3]port link-type Trunk
[SW2-Ethernet0/0/3]port Trunk allow-pass VLAN all
[SW2-Ethernet0/0/3]quit
```

（4）查看 SW1 和 SW2 上 Trunk 端口的分配情况，代码如下。

```
[SW1]display port vlan  active
T=TAG U=UNTAG
-------------------------------------------------------------------------------
Port              Link Type     PVID     VLAN List
-------------------------------------------------------------------------------
Eth0/0/1          access        10       U: 10
Eth0/0/2          access        20       U: 20
Eth0/0/3          Trunk         1        U: 1
                                         T: 10 20
Eth0/0/4          hybrid        1        U: 1
Eth0/0/5          hybrid        1        U: 1
Eth0/0/6          hybrid        1        U: 1
……

[SW2]display port vlan  active
T=TAG U=UNTAG
-------------------------------------------------------------------------------
Port              Link Type     PVID     VLAN List
-------------------------------------------------------------------------------
Eth0/0/1          access        10       U: 10
Eth0/0/2          access        20       U: 20
Eth0/0/3          Trunk         1        U: 1
                                         T: 10 20
Eth0/0/4          hybrid        1        U: 1
Eth0/0/5          hybrid        1        U: 1
```

```
Eth0/0/6              hybrid      1       U: 1
......
```

（5）测试 VLAN。打开 PC1 的命令行窗口，分别测试 PC1 到 PC3、PC1 到 PC4 是否处于连通状态。在没有划分 VLAN 10、VLAN 20 之前，它们之间都是连通状态；划分 VLAN 并设置 Trunk 后，可发现 PC1 到 PC3 处于连通状态，而 PC1 到 PC4 处于断开状态。

```
PC1>ping 10.1.1.3

Ping 10.1.1.3: 32 data bytes, Press Ctrl_C to break
From 10.1.1.3: bytes=32 seq=1 ttl=128 time=63 ms
From 10.1.1.3: bytes=32 seq=2 ttl=128 time=62 ms
From 10.1.1.3: bytes=32 seq=3 ttl=128 time=62 ms
From 10.1.1.3: bytes=32 seq=4 ttl=128 time=47 ms
From 10.1.1.3: bytes=32 seq=5 ttl=128 time=63 ms

--- 10.1.1.3 ping statistics ---
  5 packet(s) transmitted
  5 packet(s) received
0.00% packet loss
  round-trip min/avg/max = 47/59/63 ms

PC1>ping 10.1.1.4

Ping 10.1.1.8: 32 data bytes, Press Ctrl_C to break
From 10.1.1.1: Destination host unreachable
From 10.1.1.1: Destination host unreachable
From 10.1.1.1: Destination host unreachable
From 10.1.1.1: Destination host unreachable
From 10.1.1.1: Destination host unreachable

--- 10.1.1.8 ping statistics ---
  5 packet(s) transmitted
  0 packet(s) received
100.00% packet loss
```

3. VLAN（Hybrid）配置实验

图 2-24 所示为 VLAN（Hybrid）配置实验的拓扑示意，在交换机上划分 VLAN 10、VLAN 20、VLAN 30，VLAN 10 所属部门为销售部，VLAN 20 所属部门为研发部，VLAN 30 所属部门为人事部。

实验要求销售部与研发部两个部门之间不能通信，人事部可以和销售部、研发部两个部门通信，设置 Hybrid 类型，详细步骤如下。

V2-5 VLAN（Hybrid）配置实验

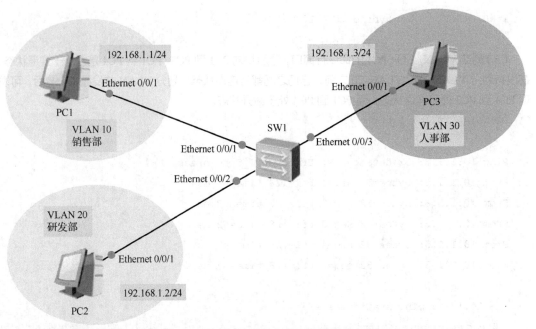

图 2-24　VLAN（Hybrid）配置实验的拓扑示意

（1）创建 VLAN 10、VLAN 20、VLAN 30，代码如下。

```
[Huawei]vlan batch 10 20 30
```

（2）将端口划入对应的 VLAN，代码如下。

```
[Huawei]int e0/0/1
[Huawei-Ethernet0/0/1]port hybrid pvid VLAN 10
[Huawei-Ethernet0/0/1]port hybrid untagged VLAN 10 30

[Huawei]int e0/0/2
[Huawei-Ethernet0/0/2]port hybrid pvid VLAN 20
[Huawei-Ethernet0/0/2]port hybrid untagged VLAN 20 30

[Huawei]int e0/0/3
[Huawei-Ethernet0/0/3]port hybrid pvid VLAN 30
[Huawei-Ethernet0/0/3]port hybrid untagged VLAN 10 20 30
```

（3）查看端口对应的 VLAN 信息，代码如下。

```
[Huawei]display port vlan active
T=TAG U=UNTAG
--------------------------------------------------------------------
Port            Link Type    PVID   VLAN List
--------------------------------------------------------------------
Eth0/0/1        hybrid       10     U: 1 10 30
Eth0/0/2        hybrid       20     U: 1 20 30
```

```
Eth0/0/3              hybrid          30          U: 1 10 20 30
Eth0/0/4              hybrid          1           U: 1
Eth0/0/5              hybrid          1           U: 1
Eth0/0/6              hybrid          1           U: 1
Eth0/0/7              hybrid          1           U: 1
Eth0/0/8              hybrid          1           U: 1
……
```

（4）测试连通性。打开销售部 PC1 的命令行窗口，分别测试其与研发部 PC2 和人事部 PC3 的连通情况，检验是否符合实验要求。在没有划分 VLAN 10、VLAN 20、VLAN 30 之前，它们之间都处于连通状态。根据实验要求设置 Hybrid 并划分 VLAN，可发现销售部与研发部处于断开状态，但它们都能与人事部连通，代码如下。

```
PC1>ping 192.168.1.2

Ping 192.168.1.2: 32 data bytes, Press Ctrl_C to break
From 192.168.1.2: Destination host unreachable
From 192.168.1.2: Destination host unreachable
From 192.168.1.2: Destination host unreachable
From 192.168.1.2: Destination host unreachable
From 192.168.1.2: Destination host unreachable

--- 192.168.1.2 ping statistics ---
  5 packet(s) transmitted
  0 packet(s) received
  100.00% packet loss

PC1>ping 192.168.1.3

Ping 192.168.1.3: 32 data bytes, Press Ctrl_C to break
From 192.168.1.3: bytes=32 seq=1 ttl=128 time=47 ms
From 192.168.1.3: bytes=32 seq=2 ttl=128 time=47 ms
From 192.168.1.3: bytes=32 seq=3 ttl=128 time=47 ms
From 192.168.1.3: bytes=32 seq=4 ttl=128 time=63 ms
From 192.168.1.3: bytes=32 seq=5 ttl=128 time=47 ms

--- 192.168.1.3 ping statistics ---
  5 packet(s) transmitted
  5 packet(s) received
  0.00% packet loss
  round-trip min/avg/max = 47/50/63 ms
```

4. GVRP 基础配置实验

GVRP 基础配置实验的拓扑示意如图 2-25 所示。

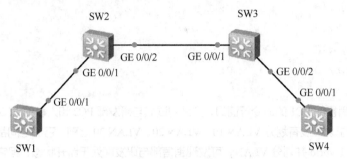

V2-6　GVRP 基础配置实验

图 2-25　GVRP 基础配置实验的拓扑示意

GVRP 配置过程如下。

（1）启用全局 GVRP，配置交换机之间的 Trunk 链路，并允许所有 VLAN 通过，启用端口 GVRP。注意：GVRP 是单向注册，所以配置 GVRP 时需要双向配置。

SW1 配置代码如下。

```
<Huawei>sys
[Huawei]sysname SW1
[SW1]GVRP //启用全局 GVRP
[SW1]int g0/0/1
[SW1-GigabitEthernet0/0/1]port link-type Trunk
[SW1-GigabitEthernet0/0/1]port Trunk allow-pass VLAN all
[SW1-GigabitEthernet0/0/1]GVRP //启用端口 GVRP
[SW1-GigabitEthernet0/0/1]GVRP registration NORMAL
/*配置 GVRP 注册模式为 NORMAL，其模式默认为 NORMAL，此命令可不执行*/
[SW1-GigabitEthernet0/0/1]quit
[SW1]
```

SW2 配置代码与 SW1 配置代码相似，简写如下。

```
sys
sysn SW2
GVRP
int g0/0/1
port link-type Trunk
port Trunk allow-pass VLAN all
GVRP
int g0/0/2
port link-type Trunk
port Trunk allow-pass VLAN all
GVRP
quit
```

SW3 配置代码与 SW1 配置代码相似，简写如下。

```
sys
sysn SW3
GVRP
int g0/0/1
port link-type Trunk
port Trunk allow-pass VLAN all
GVRP
int g0/0/2
port link-type Trunk
port Trunk allow-pass VLAN all
GVRP
quit
```

SW4 配置代码与 SW1 配置代码相似，简写如下。

```
sys
sysn SW4
GVRP
int g0/0/1
port link-type Trunk
port Trunk allow-pass VLAN all
GVRP
quit
```

（2）在 SW1 上创建 VLAN 10、VLAN 20、VLAN 30，代码如下。

```
[SW1]vlan batch 10 20 30
Info: This operation may take a few seconds. Please wait for a moment...done.
```

（3）在 SW2、SW3 和 SW4 上使用 display vlan summary 命令查看结果，可以发现在 SW2、SW3、SW4 上都能动态获取 SW1 上创建的 VLAN 10、VLAN 20、VLAN 30，代码如下。

```
[SW2]display vlan summary
static vlan:
Total 1 static vlan.
 1

dynamic vlan:
Total 3 dynamic vlan.
 10 20 30


reserved vlan:
Total 0 reserved vlan.
```

```
[SW3]display vlan summary
static vlan:
Total 1 static vlan.
 1

dynamic vlan:
Total 3 dynamic vlan.
 10 20 30

reserved vlan:
Total 0 reserved vlan.

[SW4]display vlan summary
static vlan:
Total 1 static vlan.
 1

dynamic vlan:
Total 3 dynamic vlan.
 10 20 30

reserved vlan:
Total 0 reserved vlan.
```

（4）GVRP 查看命令。使用 display GVRP status 命令验证 GVRP 的配置，查看交换机是否启用了 GVRP，或者查看 GVRP 中活动端口的信息，具体代码如下。

```
[SW3]display GVRP status
Info: GVRP is enabled
[SW3]display GVRP statistics
GVRP statistics on port GigabitEthernet0/0/1
GVRP status                : Enabled
GVRP registrations failed  : 0
GVRP last PDU origin       : 4c1f-ccb1-157b
GVRP registration type     : NORMAL
GVRP statistics on port GigabitEthernet0/0/2
GVRP status                : Enabled
GVRP registrations failed  : 0
GVRP last PDU origin       : 4c1f-ccd8-7368
GVRP registration type     : NORMAL
```

【任务拓展】

1. 配置 Access 端口

华为 VLAN 划分支持端口、MAC 地址、IP 子网、协议与策略 5 种类型，其中常用的是基于端口的 VLAN 划分。根据图 2-26 所示的 Access 端口配置拓扑示意，完成基于端口的 VLAN 划分，分别将 5 台 PC 划分到相应的 VLAN 中。

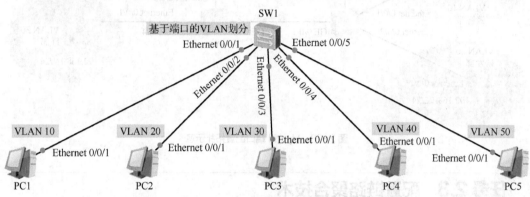

图 2-26　Access 端口配置拓扑示意

2. 配置 Trunk 端口

图 2-27 所示为 Trunk 端口配置拓扑示意。模拟某公司网络场景，该公司规模较大，有员工 200 余名，内部网络是一个大的局域网。该公司放置了多台接入交换机（如 SW1 和 SW2）负责员工的网络接入。接入交换机之间通过汇聚交换机 SW3 相连。公司通过划分 VLAN 来隔离广播域，由于员工较多，相同部门的员工通过不同交换机接入。为了保证在不同交换机下相同部门的员工能互相通信，需要配置交换机之间的链路为 Trunk 链路，以实现相同 VLAN 跨交换机通信。

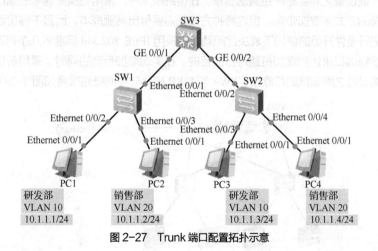

图 2-27　Trunk 端口配置拓扑示意

3. 配置 Hybrid 端口

图 2-28 所示为 Hybrid 端口配置拓扑示意。模拟企业网络，销售部、研发部、IT 部之间不能相互访问，但它们都可以和人事部进行相互访问，请完成 VLAN 划分配置。

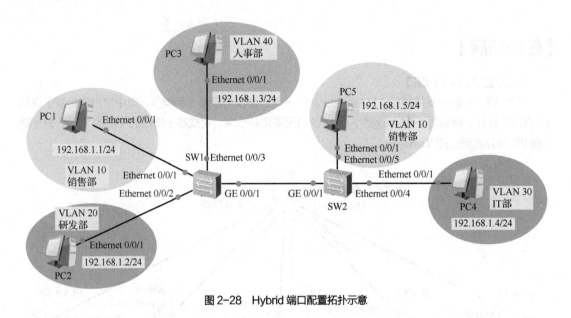

图 2-28 Hybrid 端口配置拓扑示意

任务 2.3 配置链路聚合技术

【知识准备】

2.3.1 链路聚合技术的概念和作用

随着网络规模的扩大，用户对骨干链路的带宽和可靠性提出了越来越高的要求。由于物理带宽的限制，网络的关键设备之间容易产生通信瓶颈。在传统技术中，常用更换高速率的端口板或更换高速率端口板的设备的方式来增加带宽，但这两种方式都需要付出高额成本，而且不够灵活。而链路聚合技术可以在不进行硬件升级的情况下解决此问题，它采用 IEEE 802.3ad 标准将几条物理端口捆绑在一起而形成一个逻辑端口来达到增加链路带宽的目的。在实现增加带宽的同时，采用备份链路的机制，可以有效地提高设备之间链路的可靠性。SWA 和 SWB 核心节点间链路聚合如图 2-29 所示。

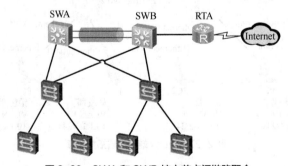

图 2-29 SWA 和 SWB 核心节点间链路聚合

1. 常见的端口类型名称

（1）发送/接收速率为 10Mbit/s 的以太网端口常被称为标准以太网端口（或标准以太口、10Mbit/s

以太网端口、10Mbit/s 以太口、10Mbit/s 口）。

（2）发送/接收速率为 100Mbit/s 的以太网端口常被称为快速以太网端口[或快速以太口、100Mbit/s 以太网端口、100Mbit/s 以太口、FE（Fast Ethernet，快速以太网）端口、FE 口]。

（3）发送/接收速率为 1000Mbit/s 的以太网端口常被称为吉比特以太网端口[或吉比特以太口、吉比特口、吉比特端口、GE（Gigabit Ethernet，吉比特以太网）端口、GE 口]。

（4）发送/接收速率为 10Gbit/s 的以太网端口常被称为万兆以太网端口（或万兆以太口、万兆口、10GE 端口、10GE 口）。

（5）发送/接收速率为 100Gbit/s 的以太网端口常被称为 100GE 端口或 100GE 口。

如果一条链路两端的端口是 GE 口，那么该链路就是一条 GE 链路；如果一条链路两端的端口是 FE 端口，那么该链路就是一条 FE 链路。

2. 链路聚合的基本概念

交换网络中通常会运行 STP，而 STP 要求设备之间不论有多少条链路，同时转发数据的链路只能有一条，因此会浪费很多冗余线路资源。当我们对网络进行链路聚合后，设备间的多条链路便可以同时工作。在企事业网络中，所有设备的流量在转发到其他网络前都会汇聚到核心层，再由核心层设备转发到其他网络，或者转发到外网。所以，在核心层设备负责数据的高速交换时，容易发生拥塞。如果在核心层部署链路聚合，则可以提升整个网络的数据吞吐量，解决拥塞问题。

链路聚合就是把两台设备之间的多条物理链路有逻辑地捆绑成一条链路，并作为一条逻辑链路来使用的技术。通过链路聚合可以提高网络带宽，并且实现成本较低。这两台进行链路聚合的设备可以是一对路由器、一对交换机或者是一台路由器和一台交换机。

一些链路聚合的相关概念如下。

（1）链路聚合组：链路聚合组（Link Aggregation Group，LAG）是指将若干条以太网链路捆绑在一起所形成的逻辑链路。

（2）链路聚合端口：每个链路聚合组唯一对应着一个逻辑端口，这个逻辑端口称为链路聚合端口，即 Eth-Trunk 端口。Eth-Trunk 端口可以作为普通以太网端口来使用，其与普通以太网端口的差别在于：转发时，链路聚合组需要从成员端口中选择一个或多个端口来进行数据转发。

（3）成员端口和成员链路：组成 Eth-Trunk 端口的各个物理端口称为成员端口，成员端口对应的链路称为成员链路。

（4）活动端口和非活动端口，以及活动链路和非活动链路：链路聚合组的成员端口存在活动端口和非活动端口两种，转发数据的端口称为活动端口，不转发数据的端口称为非活动端口；活动端口对应的链路称为活动链路，非活动端口对应的链路称为非活动链路。

（5）活动端口数上限阈值：设置活动端口数上限阈值的目的是在保证带宽的情况下提高网络的可靠性。在当前活动链路数达到上限阈值后，再向 Eth-Trunk 端口中添加成员端口时，Eth-Trunk 活动端口的数目不会继续增加，超过上限阈值的链路状态将被置为 Down。注意：手动负载均衡模式链路聚合不支持活动端口数上限阈值的配置。

（6）活动端口数下限阈值：设置活动端口数下限阈值是为了保证最小带宽。在当前活动链路数小于下限阈值时，Eth-Trunk 端口的状态转为 Down。

3. 链路聚合的主要作用

链路聚合的主要作用如下。

（1）提高网络带宽。理论上，通过聚合几条链路，一个聚合端口的带宽可以扩展为所有成员端口带宽的总和，这样就有效地增加了逻辑链路的带宽。

（2）保证网络可靠。配置链路聚合后，如果一个成员端口发生故障，则该成员端口的成员链路会把流量切换到另一条成员链路上。

（3）负载均衡。一个聚合端口可以把流量分散到多个不同的成员端口上，通过成员链路把流量发送到一个目的地，将网络产生拥塞的可能性降到最低。

2.3.2 链路聚合技术的应用场景

链路聚合技术不仅可以应用在交换机之间，还可以应用在交换机与路由器之间、路由器与路由器之间、交换机与服务器之间、路由器与服务器之间、服务器与服务器之间，如图 2-30 所示。从原理上而言，PC 上也是可以实现链路聚合的，但是成本较高，所以现实中没有真正实现。另外，从原理角度来看，服务器只是高性能的计算机，但从网络应用角度来看，服务器的地位是非常重要的，必须保证服务器与其他设备之间的连接具有非常高的可靠性。因此，服务器上经常需要用到链路聚合技术。

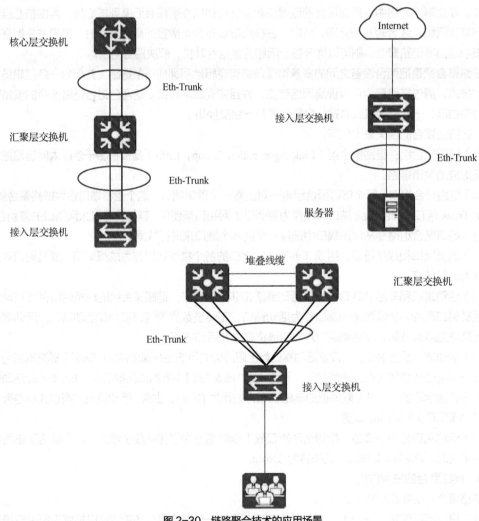

图 2-30　链路聚合技术的应用场景

2.3.3 链路聚合技术的模式

链路聚合技术的模式有两种，即手动负载均衡模式和静态链路聚合控制协议（Link Aggregation Control Protocol，LACP）模式，如图2-31所示。

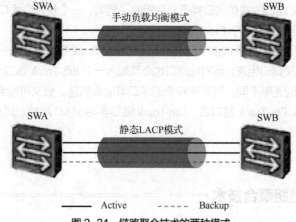

图2-31 链路聚合技术的两种模式

1. 手动负载均衡模式

在手动负载均衡模式中，Eth-Trunk端口的建立、成员端口的加入等需手动配置，没有LACP的参与。该模式下的所有活动链路都参与数据的转发，平均分担流量，因此被称为手动负载均衡模式（或手动负载分担模式）。

如果某条活动链路出现故障，则链路聚合组自动在剩余的活动链路中平均分担流量。当需要在两个直连设备之间提供一个较大的链路带宽而设备不支持LACP时，可以采用手动负载均衡模式。ARG3系列路由器和X7系列交换机可以基于目的MAC地址、源MAC地址，或者基于源MAC地址和目的MAC地址，或者基于源IP地址和目的IP地址进行负载均衡。

2. 静态LACP模式

在静态LACP模式中，链路两端的设备相互发送LACP报文，协商聚合参数。协商完成后，两台设备确定活动端口和非活动端口。在静态LACP模式中，需要手动创建一个Eth-Trunk端口，并为其添加成员端口。LACP协商选举活动端口和非活动端口。静态模式也叫$M:N$模式。其中，M代表活动链路，用于在负载均衡中转发数据；N代表非活动链路，用于冗余备份。如果一条活动链路发生故障，则该链路传输的数据将切换到一条优先级高的备份链路上，这条备份链路将转变为活动状态。

这两种模式的主要区别如下：在静态LACP模式中，一些链路充当备份链路；在手动负载均衡模式中，所有成员链路都处于转发状态。

2.3.4 链路聚合技术需要注意的规则

配置Eth-Trunk端口和成员端口时，需要注意以下规则。

（1）只能删除不包含任何成员端口的Eth-Trunk端口。

（2）把端口加入Eth-Trunk端口时，二层Eth-Trunk端口的成员端口必须是二层端口，三层Eth-Trunk端口的成员端口必须是三层端口。

（3）一个Eth-Trunk端口最多可以加入8个成员端口。

（4）加入Eth-Trunk端口的端口必须是Hybrid端口（默认的端口类型）。

（5）一个Eth-Trunk端口不能充当其他Eth-Trunk端口的成员端口。

（6）一个以太网端口只能加入一个Eth-Trunk端口。如果把一个以太网端口加入另一个Eth-Trunk端口，则必须先把该以太网端口从当前所属的Eth-Trunk端口中删除。

（7）一个Eth-Trunk端口的成员端口类型必须相同。例如，一个快速以太口（FE口）和一个吉比特以太口（GE口）不能加入同一个Eth-Trunk端口。

（8）位于不同端口板上的以太网端口可以加入同一个Eth-Trunk端口。如果一个对端端口直接和本端Eth-Trunk端口的一个成员端口相连，该对端端口也必须加入一个Eth-Trunk端口，否则两端无法通信。

（9）如果成员端口的速率不同，则速率较低的端口可能会拥塞，报文可能会被丢弃。

（10）某个端口加入Eth-Trunk端口后，Eth-Trunk端口学习MAC地址，成员端口不再学习。

【任务实施】

实验3　配置链路聚合技术

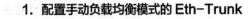

实验目的：

（1）熟悉链路聚合的工作原理。

（2）掌握链路聚合的方式。

（3）熟悉链路聚合的配置。

本实验模拟企业网络环境，SW1和SW2为企业核心交换机，分别连接两个部门。根据企业规划，SW1和SW2之间原本通过一条光纤线路相连，但从带宽和冗余角度考虑需要对其进行升级，可使用Eth-Trunk端口实现此需求。

V2-7　配置手动负载均衡模式的Eth-Trunk

1. 配置手动负载均衡模式的Eth-Trunk

手动配置负载均衡模式的Eth-Trunk的拓扑示意如图2-32所示。

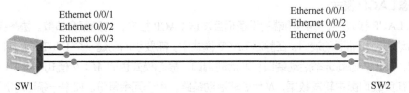

图2-32　手动配置负载均衡模式的Eth-Trunk的拓扑示意

（1）配置前查看STP。交换网络中默认运行STP，查看STP可发现SW1的3个端口都处于转发状态（FORWARDING），而SW2只有一个端口处于转发状态，另两个端口处于关闭状态（DISCARDING），具体代码如下。

```
[SW1]display stp brief
 MSTID  Port              Role  STP State    Protection
   0    Ethernet0/0/1     DESI  FORWARDING   NONE
   0    Ethernet0/0/2     DESI  FORWARDING   NONE
   0    Ethernet0/0/3     DESI  FORWARDING   NONE
```

```
[SW2]display stp brief
MSTID   Port                        Role    STP State       Protection
0       Ethernet0/0/1               ROOT    FORWARDING      NONE
0       Ethernet0/0/2               ALTE    DISCARDING      NONE
0       Ethernet0/0/3               ALTE    DISCARDING      NONE
```

（2）配置手动工负载均衡模式的 Eth-Trunk，代码如下。

```
[SW1]int Eth-Trunk 12           //创建链路聚合后的端口 12
[SW1-Eth-Trunk12]Trunkport Ethernet 0/0/1 to 0/0/3        //添加成员端口到聚合端口
Info: This operation may take a few seconds. Please wait for a moment...done.
[SW1-Eth-Trunk12]
Sep 14 2021 12:47:08-08:00 SW1 %%01IFNET/4/IF_STATE(l)[0]:Interface Eth-Trunk12
has turned into UP state.
Sep 14 2021 12:47:08-08:00 SW1 DS/4/DATASYNC_CFGCHANGE:OID 1.3.6.1.4.1.2011.5.25
.191.3.1 configurations have been changed. The current change number is 5, the c
hange loop count is 0, and the maximum number of records is 4095.
[SW1-Eth-Trunk12]quit

[SW2]int Eth-Trunk 12
[SW2-Eth-Trunk12]Trunkport Ethernet 0/0/1 to 0/0/3
Info: This operation may take a few seconds. Please wait for a moment...done.
[SW2-Eth-Trunk12]
[SW2-Eth-Trunk12]
Sep 14 2021 12:49:39-08:00 SW2 DS/4/DATASYNC_CFGCHANGE:OID 1.3.6.1.4.1.2011.5.25
.191.3.1 configurations have been changed. The current change number is 5, the c
hange loop count is 0, and the maximum number of records is 4095.
[SW2-Eth-Trunk12]quit
```

（3）查看聚合信息。手动配置后，查看 STP 可发现 SW1、SW2 的 3 个端口都合并成 Eth-Trunk 12 端口，且处于转发状态，具体代码如下。

```
[SW1]display interface brief
PHY: Physical
*down: administratively down
(l): loopback
(s): spoofing
(b): BFD down
(e): ETHOAM down
(dl): DLDP down
(d): Dampening Suppressed
InUti/OutUti: input utility/output utility
Interface               PHY     Protocol    InUti   OutUti      inErrors outErrors
Eth-Trunk12             up      up          0%      0%          0        0
```

```
    Ethernet0/0/1           up     up      0%    0%     0        0
    Ethernet0/0/2           up     up      0%    0%     0        0
    Ethernet0/0/3           up     up      0%    0%     0        0
......
[SW2]display interface brief
PHY: Physical
*down: administratively down
(l): loopback
(s): spoofing
(b): BFD down
(e): ETHOAM down
(dl): DLDP down
(d): Dampening Suppressed
InUti/OutUti: input utility/output utility
Interface              PHY    Protocol  InUti OutUti  inErrors outErrors
Eth-Trunk12             up     up        0%    0%      0        0
Ethernet0/0/1           up     up        0%    0%      0        0
Ethernet0/0/2           up     up        0%    0%      0        0
Ethernet0/0/3           up     up        0%    0%      0        0
......

[SW1]display stp brief
 MSTID   Port            Role    STP State    Protection
   0     Eth-Thunk12     DESI    FORWARDING   NONE

[SW2]display stp brief
 MSTID   Port            Role    STP State    Protection
   0     Eth-Thunk12     ROOT    FORWARDING   NONE

[SW1]display interface Eth-Trunk 12
Eth-Trunk12 current state : UP
Line protocol current state : UP
Description:
Switch Port, PVID :   1, Hash arithmetic : According to SIP-XOR-DIP, Maximal BW:
 300M, Current BW: 300M, The Maximum Frame Length is 9216
IP Sending Frames' Format is PKTFMT_ETHNT_2, Hardware address is 4c1f-cc5c-2de2
Current system time: 2021-09-14 13:00:10-08:00
    Input bandwidth utilization  :   0%
    Output bandwidth utilization :   0%
--------------------------------------------------------
```

```
PortName                         Status        Weight
--------------------------------------------------------
Ethernet0/0/1                    UP            1
Ethernet0/0/2                    UP            1
Ethernet0/0/3                    UP            1
--------------------------------------------------------
The Number of Ports in Trunk : 3
The Number of UP Ports in Trunk : 3

[SW2]display interface Eth-Trunk 12
Eth-Trunk12 current state : UP
Line protocol current state : UP
Description:
Switch Port, PVID :    1, Hash arithmetic : According to SIP-XOR-DIP, Maximal BW:
 300M, Current BW: 300M, The Maximum Frame Length is 9216
IP Sending Frames' Format is PKTFMT_ETHNT_2, Hardware address is 4c1f-cc7e-30c3
Current system time: 2021-09-14 12:58:53-08:00
    Input bandwidth utilization  :    0%
    Output bandwidth utilization :    0%
--------------------------------------------------------
PortName                         Status        Weight
--------------------------------------------------------
Ethernet0/0/1                    UP            1
Ethernet0/0/2                    UP            1
Ethernet0/0/3                    UP            1
--------------------------------------------------------
The Number of Ports in Trunk : 3
The Number of UP Ports in Trunk : 3
```

2. 配置静态 LACP 模式的 Eth-Trunk

图 2-33 所示为静态 LACP 模式 Eth-Trunk 的拓扑示意，实验要求将交换机之间的端口 Ethernet 0/0/1 和 Ethernet 0/0/2 设置为主链路（ACTIVE），端口 Ethernet 0/0/3 设置为备份链路（BACKUP），具体实现过程如下。

V2-8　配置静态 LACP 模式的 Eth-Trunk

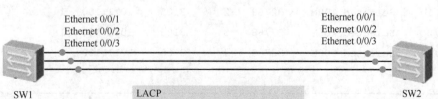

图 2-33　静态 LACP 模式 Eth-Trunk 的拓扑示意

（1）配置前查看 STP，代码如下。

```
[SW1]display stp brief
 MSTID   Port              Role   STP State    Protection
   0     Ethernet0/0/1     DESI   FORWARDING   NONE
   0     Ethernet0/0/2     DESI   FORWARDING   NONE
   0     Ethernet0/0/3     DESI   FORWARDING   NONE
[SW2]display stp brief
 MSTID   Port              Role   STP State    Protection
   0     Ethernet0/0/1     ROOT   FORWARDING   NONE
   0     Ethernet0/0/2     ALTE   DISCARDING   NONE
   0     Ethernet0/0/3     ALTE   DISCARDING   NONE
```

（2）配置静态 LACP 模式的 Eth-Trunk，代码如下。

```
[SW1]int Eth-Trunk 12
[SW1-Eth-Trunk12]mode lacp-static   //启用静态 LACP 模式
[SW1-Eth-Trunk12]Trunkport Ethernet 0/0/1 to 0/0/3
[SW1-Eth-Trunk12]quit


[SW2]int Eth-Trunk 12
[SW2-Eth-Trunk12]mode lacp-static
[SW2-Eth-Trunk12]Trunkport Ethernet 0/0/1 to 0/0/3
[SW2-Eth-Trunk12]quit
```

（3）查看 Eth-Trunk 信息，代码如下。

```
[SW1]display eth-Trunk 12
Eth-Trunk12's state information is:
Local:
LAG ID: 12                  WorkingMode: STATIC
Preempt Delay: Disabled     Hash arithmetic: According to SIP-XOR-DIP
System Priority: 32768 System ID: 4c1f-cc5c-2de2
Least Active-linknumber: 1 Max Active-linknumber: 8
Operate status: up          Number Of Up Port In Trunk: 3
--------------------------------------------------------------------
ActorPortName     Status       PortType  PortPri  PortNo  PortKey  PortState  Weight
Ethernet0/0/1     Selected     100M      32768     2       3105    10111100    1
Ethernet0/0/2     Selected     100M      32768     3       3105    10111100    1
Ethernet0/0/3     Selected     100M      32768     4       3105    10111100    1
Partner:
--------------------------------------------------------------------
ActorPortName     SysPri   SystemID            PortPri  PortNo  PortKey  PortState
Ethernet0/0/1     32768    4c1f-cc7e-30c3      32768     2       3105    10111100
Ethernet0/0/2     32768    4c1f-cc7e-30c3      32768     3       3105    10111100
```

```
Ethernet0/0/3      32768       4c1f-cc7e-30c3   32768      4          3105       10111100
```

```
[SW2]display eth-Trunk 12
Eth-Trunk12's state information is:
Local:
LAG ID: 12                    WorkingMode: STATIC
Preempt Delay: Disabled       Hash arithmetic: According to SIP-XOR-DIP
System Priority: 32768  System ID: 4c1f-cc7e-30c3
Least Active-linknumber: 1  Max Active-linknumber: 8
Operate status: up         Number Of Up Port In Trunk: 3
--------------------------------------------------------------------------
ActorPortName    Status    PortType  PortPri  PortNo  PortKey  PortState  Weight
Ethernet0/0/1    Selected  100M      32768    2       3105     10111100   1
Ethernet0/0/2    Selected  100M      32768    3       3105     10111100   1
Ethernet0/0/3    Selected  100M      32768    4       3105     10111100   1
Partner:
--------------------------------------------------------------------------
ActorPortName    SysPri    SystemID              PortPri  PortNo  PortKey  PortState
Ethernet0/0/1    32768     4c1f-cc5c-2de2        32768    2       3105     10111100
Ethernet0/0/2    32768     4c1f-cc5c-2de2        32768    3       3105     10111100
Ethernet0/0/3    32768     4c1f-cc5c-2de2        32768    4       3105     10111100
```

通过使用 display eth-Trunk 12 命令查看可知，两台设备的系统优先级（System Priority）都为32768，同时可以看到两台设备的3个端口都处于活动状态（Selected），端口优先级（PortPri）都为32768。设备优先级与端口优先级可以通过改变该参数的大小而修改，该参数的值越小表示优先级越高。同时可以看到3条链路都处于活动状态（Selected），最大聚合线路（Max Active-linknumber）为8条，各子端口的带宽（PortType）为100Mbit/s。

3. 主链路与备份链路

（1）可以通过修改设备的优先级大小，决定哪台交换机成为主设备，其中优先级最小的设备将成为主设备，它可以决定哪条链路是主链路，哪条链路是备份链路。具体代码如下。

```
[SW1]lacp priority 100         //设置LACP的主设备为SW1
[SW1]int Ethernet 0/0/1
[SW1-Ethernet0/0/1]lacp priority 100   //设置LACP端口优先级
[SW1]int Ethernet 0/0/2
[SW1-Ethernet0/0/2]lacp priority 100
[SW1-Ethernet0/0/2]quit
[SW1]dis eth-Trunk 12          //查看参数改变
Eth-Trunk12's state information is:
Local:
LAG ID: 12                    WorkingMode: STATIC
Preempt Delay: Disabled       Hash arithmetic: According to SIP-XOR-DIP
```

```
System Priority: 100        System ID: 4c1f-cc5c-2de2
Least Active-linknumber: 1  Max Active-linknumber: 8
Operate status: up          Number Of Up Port In Trunk: 3
--------------------------------------------------------------------------
ActorPortName    Status    PortType   PortPri  PortNo  PortKey  PortState  Weight
Ethernet0/0/1    Selected  100M       100      2       3105     10111100   1
Ethernet0/0/2    Selected  100M       100      3       3105     10111100   1
Ethernet0/0/3    Selected  100M       32768    4       3105     10111100   1
Partner:
--------------------------------------------------------------------------
ActorPortName    SysPri    SystemID             PortPri  PortNo  PortKey  PortState
Ethernet0/0/1    32768     4c1f-cc7e-30c3       32768    2       3105     10111100
Ethernet0/0/2    32768     4c1f-cc7e-30c3       32768    3       3105     10111100
Ethernet0/0/3    32768     4c1f-cc7e-30c3       32768    4       3105     10111100
```

（2）当对设备和端口优先级进行调整后，3 条链路还是处于活动状态（Selected），修改最大活动链路数量为 2，代码如下。

```
[SW1]int Eth-Trunk 12
[SW1-Eth-Trunk12]max active-linknumber 2  //设置最大活动链路数量
[SW1]dis eth-Trunk 12
Eth-Trunk12's state information is:
Local:
LAG ID: 12                  WorkingMode: STATIC
Preempt Delay: Disabled     Hash arithmetic: According to SIP-XOR-DIP
System Priority: 100        System ID: 4c1f-cc5c-2de2
Least Active-linknumber: 1  Max Active-linknumber: 2
Operate status: up          Number Of Up Port In Trunk: 2
--------------------------------------------------------------------------
ActorPortName    Status    PortType   PortPri  PortNo  PortKey  PortState  Weight
Ethernet0/0/1    Selected  100M       100      2       3105     10111100   1
Ethernet0/0/2    Selected  100M       100      3       3105     10111100   1
Ethernet0/0/3    Unselect  100M       32768    4       3105     10100000   1
Partner:
--------------------------------------------------------------------------
ActorPortName    SysPri    SystemID             PortPri  PortNo  PortKey  PortState
Ethernet0/0/1    32768     4c1f-cc7e-30c3       32768    2       3105     10111100
Ethernet0/0/2    32768     4c1f-cc7e-30c3       32768    3       3105     10111100
Ethernet0/0/3    32768     4c1f-cc7e-30c3       32768    4       3105     10110000
```

（3）此时，可以发现 Ethernet 0/0/3 链路没有被选择（Unselect），处于备份状态；Ethernet 0/0/1 和 Ethernet 0/0/2 处于活动状态，成为主链路。代码如下。

```
[SW1]dis int Eth-Trunk 12
Eth-Trunk12 current state : UP
```

```
Line protocol current state : UP
Description:
Switch Port, PVID :    1, Hash arithmetic : According to SIP-XOR-DIP, Maximal BW:
 300M, Current BW: 200M, The Maximum Frame Length is 9216
IP Sending Frames' Format is PKTFMT_ETHNT_2, Hardware address is 4c1f-cc5c-2de2
Current system time: 2021-09-14 13:32:54-08:00
Input bandwidth utilization :    0%
Output bandwidth utilization :   0%
--------------------------------------------------
PortName                  Status      Weight
--------------------------------------------------
Ethernet0/0/1             UP          1
Ethernet0/0/2             UP          1
Ethernet0/0/3             DOWN        1
--------------------------------------------------
The Number of Ports in Trunk : 3
The Number of UP Ports in Trunk : 2
```

（4）查看 Eth-Trunk 12 端口带宽，最大带宽为 300Mbit/s，当前带宽为 200Mbit/s。

若主链路某个端口失效，则查看备用链路是否被激活。假设关闭主链路 Ethernet 0/0/1 后，Ethernet 0/0/2 和 Ethernet 0/0/3 被选择成为主链路，Ethernet 0/0/1 没有被选择，成为备用链路。具体代码如下。

```
[SW1]int e0/0/1
[SW1-Ethernet0/0/1]shutdown  //关闭端口
[SW1]display eth-Trunk 12
Eth-Trunk12's state information is:
Local:
LAG ID: 12                  WorkingMode: STATIC
Preempt Delay: Disabled     Hash arithmetic: According to SIP-XOR-DIP
System Priority: 100        System ID: 4c1f-cc5c-2de2
Least Active-linknumber: 1  Max Active-linknumber: 2
Operate status: up          Number Of Up Port In Trunk: 2
--------------------------------------------------------------------------------
ActorPortName   Status     PortType  PortPri  PortNo   PortKey  PortState  Weight
Ethernet0/0/1   Unselect   100M      100      2        3105     10100010   1
Ethernet0/0/2   Selected   100M      100      3        3105     10111100   1
Ethernet0/0/3   Selected   100M      32768    4        3105     10111100   1
Partner:
--------------------------------------------------------------------------------
ActorPortName   SysPri   SystemID          PortPri   PortNo   PortKey   PortState
Ethernet0/0/1   0        0000-0000-0000    0         0        0         10100011
Ethernet0/0/2   32768    4c1f-cc7e-30c3    32768     3        3105      10111100
```

```
Ethernet0/0/3    32768  4c1f-cc7e-30c3  32768     4     3105    10111100
```

（5）重新开启端口 1，启用 LACP 抢占功能，经过 30s，此时 Ethernet 0/0/1 和 Ethernet 0/0/2 被选择成为主链路，Ethernet 0/0/3 没有被选择，成为备用链路。具体代码如下。

```
[SW1]interface e0/0/1
[SW1-Ethernet0/0/1]undo shutdown           //开启端口
[SW1-Ethernet0/0/1]quit
[SW1]int Eth-Trunk 12
[SW1-Eth-Trunk12]lacp preempt enable       //启用 LACP 抢占功能，经过 30s 后才有效
[SW1]display eth-Trunk 12
```

【任务拓展】

链路聚合拓扑示意如图 2-34 所示，分别使用两种链路聚合模式完成实验配置，完成 STP 参数修改，使 PC1 与 PC2 间的通信必须通过主链路。

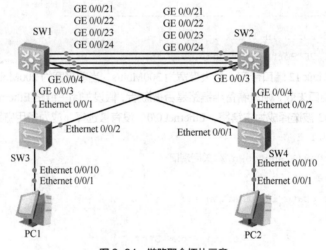

图 2-34　链路聚合拓扑示意

模块 3
路由器互联技术

随着宽带接入的普及,很多家庭和小型企业都组建了局域网来共享宽带接入,但是随着局域网规模的扩大,很多应用场景涉及 2 台或以上的路由器。当一个局域网内存在 2 台及以上的路由器时,不同主机间常有互访的需求,这就需要在局域网内设置路由。网络规模较小且不经常变动时,手动配置静态路由即可使得数据包能够按照预定的路径传送到指定的目的网络。但是在大规模网络中,由于静态路由不能随着网络的变化自动调整,人工管理路由器的路由表是一项非常艰巨的任务且容易出错。而动态路由能轻松、自动地构建路由表,完成路径的最佳选择与转发工作。本模块主要介绍路由器互联技术,包括静态路由的配置和动态路由的配置。

【学习目标】

知识目标
(1)理解路由的概念及路由表相关参数信息。
(2)理解 IP 地址的结构与类型。
(3)理解网络层的作用及路由器的工作原理。
(4)熟悉常见的几种静态路由的概念。
(5)理解 RIP 路由协议的基本概念及 RIP 路由的工作原理。
(6)理解 OSPF 路由协议的基本概念及 OSPF 路由的工作原理。
(7)理解 DR 和 BDR 的选举过程及 OSPF 区域划分。
(8)理解三层交换机的三层端口的作用。

技能目标
(1)掌握静态路由与默认路由的配置方法。
(2)掌握 RIP 动态路由的配置方法。
(3)掌握 OSPF 单区域与多区域动态路由的配置方法。
(4)熟悉 OSPF 链路认证与区域认证的配置方法。
(5)熟悉单臂路由与 VLAN 间路由的配置方法。

素质目标
(1)培养学生树立正确的目标与高远志向,并为之不懈奋斗、自强不息。
(2)培养学生的社会责任感,为我国的科技发展贡献自己的一份力量。

任务 3.1 静态路由的配置

【知识准备】

3.1.1 虚电路与数据报

网络层提供了虚电路与数据报两种服务方式。

1. 虚电路

虚电路（Virtual Circuit，VC）是面向有连接的数据传输，工作过程类似于线路交换，不同之处在于此时的电路是虚拟的。虚电路只是一条逻辑上的连接，分组沿着这条逻辑连接按照存储转发方式传送，而不是真正建立了一条物理连接。采用虚电路方式传输时，物理介质被理解为由多个子信道[称为逻辑信道（Logical Channel，LC）]组成，子信道的串接形成虚电路，利用不同的虚电路可支持不同的用户数据的传输。虚电路服务如图 3-1 所示。

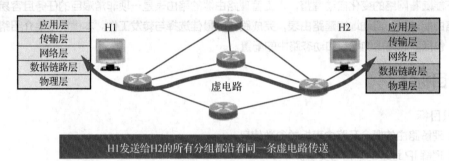

图 3-1 虚电路服务

采用虚电路进行数据传输的过程如下。

（1）虚电路建立：发送方发送含有地址信息的特定的控制信息块（如呼叫分组），该信息块途经的每个中间节点都根据当前的逻辑信道使用状况分配逻辑信道，并建立输入/输出逻辑信道映射表，所有中间节点分配的逻辑信道的串接形成虚电路。

（2）数据传输：节点发送的所有分组均沿着相同的虚电路传输，分组的收发顺序完全相同。

（3）虚电路释放：数据传输完毕后，采用特定的控制信息块（如拆除分组）释放该虚电路。通信的双方都可发起释放虚电路的动作。

由于虚电路的建立和释放需要占用一定的时间，因此虚电路方式不适用于节点之间具有频繁连接和交换短小数据的应用，如交互式的通信。

2. 数据报

网络层向上只提供简单且灵活的、无连接的、尽最大努力交付的数据报服务，不提供服务质量的承诺。数据报服务如图 3-2 所示。

由于传输网络不提供端到端的可靠传输服务，这就使网络中的路由器可以做得比较简单，而且价格低廉，如果主机（即端系统）中的进程之间的通信需要是可靠的，那么由网络的主机中的传输层负责可靠交付（包括差错处理、流量控制等）。采用这种设计思路的好处如下：网络的造价大大降低，运行方式灵活，能够适应多种应用。

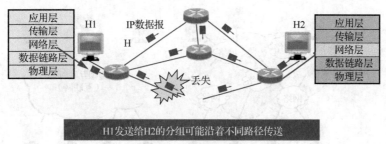

图 3-2 数据报服务

3.1.2 IP

IP 是 TCP/IP 体系中的两种主要的协议之一，也是重要的互联网标准协议之一，可以使互相连接的许多计算机网络进行通信，主要有两个版本，即 IPv4 和 IPv6，本书主要介绍 IPv4。与 IP 配套使用的还有地址解析协议（Address Resolution Protocol，ARP）、互联网控制报文协议（Internet Control Message Protocol，ICMP）、互联网组管理协议（Internet Group Management Protocol，IGMP）等。TCP/IP 体系中各层的主要协议如图 3-3 所示。

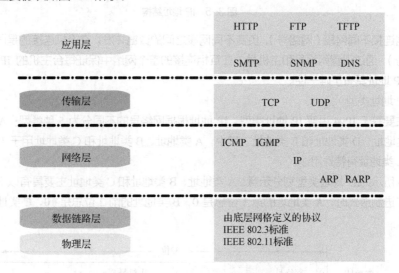

图 3-3 TCP/IP 体系中各层的主要协议

1. 互联网络与虚拟互联网络

互联网络是指通过路由器进行网络互联和路由选择；虚拟互联网络就是逻辑互联网络，利用 IP 使这些性能各异的网络从用户端看起来就像是一个统一的网络，如图 3-4 所示。

2. IP 地址

IP 地址是 IP 提供的一种统一的地址格式，它为互联网上的每一个网络和每一台主机分配一个逻辑地址，以此来屏蔽物理地址的差异。

常见的 IP 地址分为 IPv4 地址与 IPv6 地址两大类，当前广泛应用的是 IPv4 地址。目前 IPv4 地址几乎耗尽，下一阶段必然会升级到 IPv6 地址。IP 地址对应于 OSI 参考模型的网络层，工作在网络层的路由器根据目 IP 地址和源 IP 地址来判断是否属于同一网段。如果属于不同网段，则转发数据包。

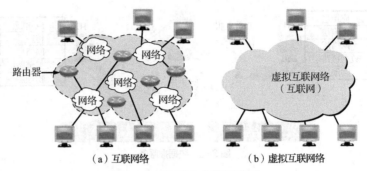

图 3-4 互联网络与虚拟互联网络

（1）IP 地址结构

计算机的 IP 地址由两部分组成，如图 3-5 所示，一部分为网络号，另一部分为主机号，同一网段内的计算机网络号部分相同，主机号部分不能同时重复出现。

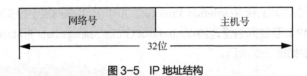

图 3-5 IP 地址结构

路由器连接不同网段（网络号），负责不同网段之间的数据转发。交换机连接的是同一网段的计算机（主机号）。通过设置网络号和主机号，在互相连接的整个网络中保证每台主机的 IP 地址不会互相重叠，即 IP 地址具有了唯一性。

（2）IP 地址类型

这里主要基于 IPv4 讲解 IP 地址类型。IP 地址根据网络号的不同分为 5 种类型：A 类地址、B 类地址、C 类地址、D 类地址和 E 类地址。其中，A 类地址、B 类地址和 C 类地址用于 TCP/IP 节点，D 类地址和 E 类地址有特殊用途。

图 3-6 所示为 IP 地址类型划分示意。A 类地址、B 类地址和 C 类地址主要具有以下特征：当将 IP 地址写作二进制形式时，A 类地址的第 1 位总是 0，B 类地址的前 2 位总是 10，C 类地址的前 3 位总是 110。

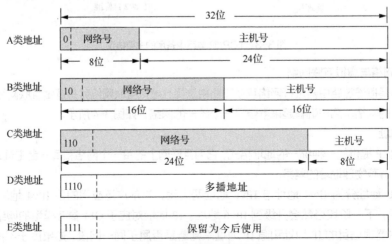

图 3-6 IP 地址类型划分示意

① A 类地址

一个 A 类地址由 1 字节的网络号和 3 字节的主机号组成，网络号的最高位必须是 0。可用的 A 类网络有 2^7-2 个，每个网络能容纳 $2^{24}-2$ 台主机。

A 类地址范围为 1.0.0.1～126.255.255.254。

A 类地址中包括私有 ID 和保留 ID。所谓的私有 ID 就是在互联网上不使用，而被用在局域网中的地址。10.x.x.x 是私有 ID；127.x.x.x 是保留 ID，用于循环测试。

② B 类地址

一个 B 类地址由 2 字节的网络号和 2 字节的主机号组成，网络号的最高位必须是 10。可用的 B 类网络有 2^{14} 个，每个网络能容纳 $2^{16}-2$ 台主机。

B 类地址范围为 128.0.0.1～191.255.255.255。

B 类地址中包括私有 ID 和保留 ID。172.16.0.0～172.31.255.255 是私有 ID，169.254.x.x 是保留 ID。如果 IP 地址是自动获取的，而在网络上又没有找到可用的 DHCP 服务器，就会得到保留 ID 中的一个 IP 地址。

③ C 类地址

C 类地址由 3 字节的网络号和 1 字节的主机号组成，网络号的最高位必须是 110。C 类网络可达 2^{21} 个，每个网络能容纳 2^8-2 台主机。

C 类地址范围为 192.0.0.1～223.255.255.254。

C 类地址中的 192.168.x.x 是私有 ID。

④ D 类地址

D 类地址不区分网络号和主机号，以 1110 开始，它是一个专门保留的地址。它并不指向特定的网络，目前这一类地址被用在多播（Multicast）中。多播地址用来标识共享同一协议的一组计算机。

D 类地址范围为 224.0.0.1～239.255.255.254。

⑤ E 类地址

E 类地址也不区分网络号和主机号，以 1111 开始，为将来使用保留。全 0 地址（0.0.0.0）对应为当前主机。全 1 地址（255.255.255.255）是当前子网的广播地址。

E 类地址范围为 240.0.0.1～247.255.255.254。

3. 子网掩码

子网掩码又称为网络掩码、地址掩码等，是一个应用于 TCP/IP 网络的 32 位二进制数。子网掩码可以屏蔽 IP 地址中的一部分，从而分离出 IP 地址中的网络号与主机号。基于子网掩码，管理员可以将网络进一步划分为若干子网。子网掩码必须结合 IP 地址一起使用。

（1）子网掩码的分类

子网掩码可以分为默认子网掩码和自定义子网掩码。默认子网掩码即未划分子网，对应的网络号都置 1，主机号都置 0。

A 类网络默认子网掩码为 255.0.0.0；B 类网络默认子网掩码为 255.255.0.0；C 类网络默认子网掩码为 255.255.255.0。

自定义子网掩码可将一个网络划分为几个子网，需要每一段使用不同的网络号或子网号，实际上可以认为是将主机号分为子网号和子网主机号。自定义子网掩码形式如下。

未做子网划分时：IP 地址=网络号+主机号。

进行子网划分后：IP 地址=网络号+子网号+子网主机号。

可见，IP 地址在划分子网后，以前的主机号位置的一部分给了子网号，余下的是子网主机号。

（2）子网掩码的确定

在划分子网之前，一定要考虑网络目前的需求和将来的需求计划，主要从两个方面考虑：一是网络中物理网段的数量（即要划分的子网数量）；二是每个物理网段的主机的数量。

下面介绍如何确定物理网段的数量，将其转换为二进制数，并确定位数 n。例如，需要 6 个子网，6 的二进制数为 110，共 3 位，即 $n=3$。

按照 IP 地址的类型写出其默认子网掩码。例如，如果是 C 类地址，则默认子网掩码为 255.255.255.0，对应二进制数为 11111111.11111111.11111111.00000000。

将子网掩码中与主机号的前 n 位对应的位置 1，其余位置 0。例如，$n=3$ 时，如果是 C 类地址，则子网掩码为 11111111.11111111.11111111.11100000，对应十进制数为 255.255.255.224；如果是 B 类地址，则子网掩码为 11111111.11111111.11100000.00000000，对应十进制数为 255.255.224.0；如果是 A 类地址，则子网掩码为 11111111.11100000.00000000.00000000，对应十进制数为 255.224.0.0。

注意　由于网络被划分为 6 个子网，占用了主机号的前 3 位，若是 C 类地址，则只能用 5 位来表示主机号，因此每个子网内的主机数为 $(2^5-2)=30$，6 个子网总共所能标识的主机数将小于 254。

3.1.3　路由的概念

1. 路由与路由表

在网络通信中，路由（Routing）是网络层的一个术语，它是指从某一网络设备出发去往某个目的地的路径。网络中路由器（或三层交换机）负责为数据包选择转发路径。路由转发过程如图 3-7 所示。

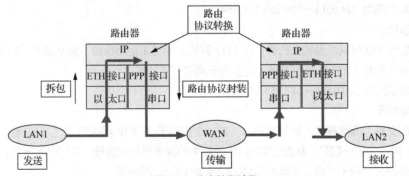

图 3-7　路由转发过程

路由器中有路由表（Routing Table），路由表是若干条路由信息的一个集合体。在路由表中，一条路由信息也被称为一个路由项或一个路由条目，路由器根据路由表为数据包选择转发路径。路由表只存在于终端计算机和路由器（以及三层交换机）中，二层交换机是不存在路由表的。

每一台具备路由功能的路由器都会维护路由表，路由表相当于路由器的地图，得益于这张地图，路由器才能够正确地转发 IP 报文。路由表中装载着路由器通过各种途径获知的路由条目，每一个路由条目包含目的网络地址、子网掩码长度、协议类型（路由的来源）、出端口、下一跳 IP 地址、优先级及度量值等信息。路由表是每台支持路由功能的路由器进行数据转发的依据和基础，是一个非常重要的概念。如图 3-8 所示，当路由器 R1 收到一个 IP 报文时，路由器会解析出报文的 IP 头部中的目的 IP 地址，然后在自己的路由表中查询该目的 IP 地址，它发现数据包的目的 IP 地址是 192.168.20.1，而路由表中存在到达 192.168.20.0/24 的路由,因此 R1 根据路由条目所指示的出端口及下一跳 IP 地址将报文转发出去。

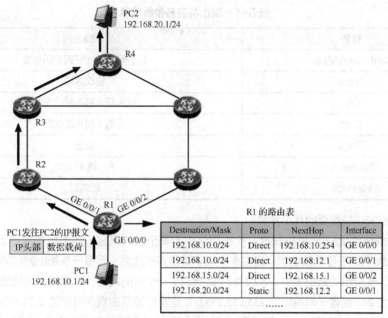

图 3-8 路由与路由表示意

2. 路由表参数

路由器的主要工作就是为经过路由器的每个数据帧寻找一条最佳传输路径,并将该数据帧有效地传送到目的站点。由此可见,选择最佳传输路径的策略(即路由算法)是路由器的关键所在。为了完成这项工作,在路由器中保存着各种传输路径的相关数据——路由表,供路由选择时使用。路由表就像平时使用的地图一样,标识着各种路线,路由表中保存着子网的标志信息、网络中路由器的数量和下一台路由器的名称等内容。

在具备路由功能的华为数据通信产品上查看路由表的命令是 display ip routing-table。图 3-9 展示了一个设备的路由表信息的示例,路由表中的每一行就是一个路由条目。在一个大型的网络中,路由器的路由表可能包含大量的路由条目。每个路由条目都采用目的网络地址及子网掩码进行标识。

```
<router>display ip routing-table
Route Flage:R-relay,D-download to fib
------------------------------------------------------------------
Routing Tables:Public
        Destinations:9        Routes:9
```

Destination/Mask	Proto	Pre	Cost	Flags	NextHop	Interface
2.2.2.0/24	Static	60	0	RD	10.1.12.2	GigabitEthernet0/0/0
10.1.12.0/24	Direct	0	0	D	10.1.12.1	GigabitEthernet0/0/0
10.1.12.1/32	Direct	0	0	D	127.0.0.1	GigabitEthernet0/0/0
22.22.22.22/32	OSPF	10	1	D	10.1.12.2	GigabitEthernet0/0/0
127.0.0.0/8	Direct	0	0	D	127.0.0.1	InLoopBack0
127.0.0.1/32	Direct	0	0	D	127.0.0.1	InLoopBack0
……						
目的网络地址及子网掩码长度	协议类型	优先级	度量值	标志	下一跳IP地址	出端口

图 3-9 查看设备的路由表信息

从路由表的输出可以看出，每个路由条目都包括多个参数信息。路由条目各参数信息如表 3-1 所示。

表 3-1 路由条目各参数信息

列名	描述
Destination/Mask	目的网络地址/子网掩码长度
Proto	协议类型
Pre	优先级（越小越优先）
Cost	度量值（越小越优先）
Flags	标志
NextHop	下一跳 IP 地址
Interface	出端口

这里只介绍其中较重要的几个参数。

（1）目的网络地址/子网掩码长度：代表网络中的某个目的网络（或者目的网段），路由表中的 Destination/Mask 列用于显示该信息。路由表相当于路由器的地图，而每一条路由都指向网络中的某个目的网络。图 3-9 所示的路由表中，2.2.2.0/24 就标识了一个目的网络，其中目的网络地址为 2.2.2.0，子网掩码长度为 24（或者子网掩码为 255.255.255.0），这意味着路由器拥有到达 2.2.2.0/24 的路由信息。

（2）协议类型：表示该路由的协议类型，路由表中的 Proto 列用于显示该信息。例如，对于 2.2.2.0/24 这条路由，Proto 列显示的是 Static，这意味着这条路由是通过手动的方式配置的静态路由。再如，对于 22.22.22.22/32 这条路由，这是一条主机路由（子网掩码为 255.255.255.255），而这条路由的 Proto 列显示的是 OSPF，表明这条路由是通过 OSPF 协议学习到的。如果 Proto 列显示 Direct，则表明这条路由为直连路由，也就是这条路由所指向的网段是设备的直连端口所在的网段。

（3）优先级：表示路由优先级，路由表中的 Pre 列用于显示该信息。路由条目的获取来源有多种，每种类型的路由对应不同的优先级，路由优先级的值越小，表示该路由的优先级越高。当一台路由器同时从多种不同的来源学习到去往同一个目的网段的路由时，它将选择优先级值最小的那条路由。路由优先级如表 3-2 所示。

表 3-2 路由优先级

路由类型	优先级
直连路由	0
OSPF 内部路由	10
IS-IS 路由	15
静态路由	60
RIP 路由	100
OSPF ASE 路由（网络外部路由）	150
OSPF NSSA 区域路由	150
IBGP 路由	255
EBGP 路由	255

（4）度量值：指示了此路由器到达目的网段的代价值，也被称为开销，路由表中的 Cost 列用于显示该信息。度量值的大小会影响到路由的优选。直连路由及静态路由默认的度量值为 0。此外，每一种动态路由协议都定义了其路由的度量值计算方法，不同的路由协议，对于路由度量值的定义和计算均有所不同。

（5）下一跳 IP 地址：描述的是路由器转发到达目的网段的数据包所使用的下一跳 IP 地址，路由表中的 NextHop 列用于显示该信息。在图 3-9 中，2.2.2.0/24 路由的 NextHop 列显示为 10.1.12.2，这意味着如果该路由器收到一个数据包，经过路由查询后发现数据包的目的地址匹配 2.2.2.0/24 这条路由，则该路由器会将数据包转发给 10.1.12.2 这个下一跳 IP 地址。

（6）出端口：指示的是数据包被路由后离开此路由器的端口，路由表中的 Interface 列用于显示该信息。还是以 2.2.2.0/24 路由为例，这条路由的 Interface 列显示的是 GigabitEthernet0/0/0，这意味着如果该路由器收到一个数据包且经过路由查询后发现数据包的目的地址匹配该路由，则该路由器会将数据包转发给 10.1.12.2 这个下一跳 IP 地址，并从 GigabitEthernet0/0/0 端口送出。

3.1.4 静态路由的概念

静态路由是指由管理员手动配置和维护的路由，由管理员事先设置好，一般在系统安装时就根据网络的配置情况预先设定，它不会随未来网络结构的改变而改变。静态路由因其配置简单、开销小而广泛应用于网络中，当一个网络发生故障后，静态路由不会自动发生改变，因而静态路由适用于网络拓扑简单的网络。

配置静态路由的命令：ip route-static <目的网络地址> <子网掩码> <下一跳 IP 地址>。静态路由配置示意如图 3-10 所示。

图 3-10 静态路由配置示意

浮动路由在网络中主路由失效的情况下，会加入路由表并承担数据转发业务。图 3-11 所示为浮动路由配置示意。

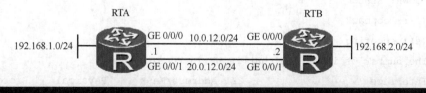

图 3-11 浮动路由配置示意

【任务实施】

实验 1　配置静态路由

实验目的：

（1）理解直连路由、静态路由、等价路由、浮动路由、默认路由的概念。

（2）掌握静态路由的配置方法。

（3）掌握等价路由、浮动路由参数的设置方法。

（4）熟悉默认路由的应用场景与配置。

假设在小型企业网络中，企业三层交换机或者路由器组网的设备数量有限，总的网络路由数量少，在维护的工作量可控的情况下，网络管理员可以考虑采用静态路由。另外，针对出口设备，如路由器、防火墙等对接公网 Internet 的默认路由都可以采用静态路由来实现。

1. 直连路由

直连路由是路由器的直连网段生成的。当路由器的端口配置了 IP 地址，并且端口处于启用状态时，路由器就会生成一条端口 IP 地址对应网段的直连路由。形象地看，直连路由就是路由器"家门口的路"，其拓扑示意如图 3-12 所示。

图 3-12　直连路由拓扑示意

V3-1　配置静态路由（直连路由、静态路由）

（1）配置端口 IP 地址，代码如下。

```
[R1]int g0/0/0
[R1-GigabitEthernet0/0/0]ip address 12.1.1.1 24


[R2]int g0/0/0
[R2-GigabitEthernet0/0/0]ip address 12.1.1.2 24
```

（2）查看端口 IP 地址，代码如下。

```
[R1]display ip interface brief
*down: administratively down
^down: standby
(l): loopback
(s): spoofing
The number of interface that is UP in Physical is 2
Interface                 IP Address/Mask      Physical        Protocol
GigabitEthernet0/0/0      12.1.1.1/24          up              up
GigabitEthernet0/0/1      unassigned           down            down
GigabitEthernet0/0/2      unassigned           down            down
NULL0                     unassigned           up              up(s)
```

```
[R2]display ip interface brief
*down: administratively down
^down: standby
(l): loopback
(s): spoofing
The number of interface that is UP in Physical is 2
Interface                   IP Address/Mask     Physical    Protocol
GigabitEthernet0/0/0        12.1.1.2/24         up          up
GigabitEthernet0/0/1        unassigned          down        down
GigabitEthernet0/0/2        unassigned          down        down
NULL0                       unassigned          up          up(s)
```

（3）查看直连路由，代码如下。

```
[R1]display ip routing-table
Route Flags: R - relay, D - download to fib
------------------------------------------------------------------------------
Routing Tables: Public
        Destinations : 7        Routes : 7
Destination/Mask    Proto   Pre  Cost  Flags NextHop    Interface
12.1.1.0/24         Direct  0    0     D     12.1.1.1   GigabitEthernet0/0/0
12.1.1.1/32         Direct  0    0     D     127.0.0.1  GigabitEthernet0/0/0
12.1.1.255/32       Direct  0    0     D     127.0.0.1  GigabitEthernet0/0/0
127.0.0.0/8         Direct  0    0     D     127.0.0.1  InLoopBack0
127.0.0.1/32        Direct  0    0     D     127.0.0.1  InLoopBack0
127.255.255.255/32  Direct  0    0     D     127.0.0.1  InLoopBack0
255.255.255.255/32  Direct  0    0     D     127.0.0.1  InLoopBack0


[R2]display ip routing-table
Route Flags: R - relay, D - download to fib
------------------------------------------------------------------------------
Routing Tables: Public
        Destinations : 7        Routes : 7
Destination/Mask    Proto   Pre  Cost  Flags NextHop    Interface
12.1.1.0/24         Direct  0    0     D     12.1.1.2   GigabitEthernet0/0/0
12.1.1.2/32         Direct  0    0     D     127.0.0.1  GigabitEthernet0/0/0
12.1.1.255/32       Direct  0    0     D     127.0.0.1  GigabitEthernet0/0/0
127.0.0.0/8         Direct  0    0     D     127.0.0.1  InLoopBack0
127.0.0.1/32        Direct  0    0     D     127.0.0.1  InLoopBack0
127.255.255.255/32  Direct  0    0     D     127.0.0.1  InLoopBack0
255.255.255.255/32  Direct  0    0     D     127.0.0.1  InLoopBack0
```

2. 静态路由

静态路由是由管理员手动添加的路由条目，通过静态路由添加的都是非直连网段。静态路由主要有两个特点：静态路由的添加和删除都需要手动完成；静态路由无法适应网络的动态变更。静态路由拓扑示意如图3-13所示。

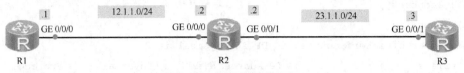

图3-13 静态路由拓扑示意

（1）配置端口IP地址，代码如下。

```
[R1]int g0/0/0
[R1-GigabitEthernet0/0/0]ip address 12.1.1.1 24

[R2]int g0/0/0
[R2-GigabitEthernet0/0/0]ip address 12.1.1.2 24
[R2]int g0/0/1
[R2-GigabitEthernet0/0/0]ip address 23.1.1.2 24

[R3]int g0/0/1
[R3-GigabitEthernet0/0/1]ip address 23.1.1.3 24
```

（2）在R1上配置到23.1.1.0/24网络的静态路由，在R3上配置到12.1.1.0/24网络的静态路由，代码如下。

```
[R1]ip route-static 23.1.1.0 24 12.1.1.2
[R3]ip route-static 12.1.1.0 24 23.1.1.2
```

（3）分别在R1、R3上查看静态路由信息，代码如下。

```
[R1]display ip routing-table
Route Flags: R - relay, D - download to fib
------------------------------------------------------------------------
Routing Tables: Public
     Destinations : 8       Routes : 8
Destination/Mask    Proto   Pre  Cost Flags NextHop    Interface
12.1.1.0/24    Direct  0     0     D    12.1.1.1  GigabitEthernet0/0/0
12.1.1.1/32    Direct  0     0     D    127.0.0.1 GigabitEthernet0/0/0
12.1.1.255/32  Direct  0     0     D    127.0.0.1 GigabitEthernet0/0/0
23.1.1.0/24    Static  60    0     RD   12.1.1.2  GigabitEthernet0/0/0
127.0.0.0/8    Direct  0     0     D    127.0.0.1   InLoopBack0
127.0.0.1/32   Direct  0     0     D    127.0.0.1   InLoopBack0
127.255.255.255/32 Direct 0  0     D    127.0.0.1   InLoopBack0
255.255.255.255/32 Direct 0  0     D    127.0.0.1   InLoopBack0

[R3]display ip routing-table
```

```
Route Flags: R - relay, D - download to fib
Routing Tables: Public
         Destinations : 8       Routes : 8
Destination/Mask    Proto  Pre  Cost  Flags NextHop     Interface
12.1.1.0/24         Static 60   0     RD    23.1.1.2    GigabitEthernet0/0/1
23.1.1.0/24         Direct 0    0     D     23.1.1.3    GigabitEthernet0/0/1
23.1.1.3/32         Direct 0    0     D     127.0.0.1   GigabitEthernet0/0/1
23.1.1.255/32       Direct 0    0     D     127.0.0.1   GigabitEthernet0/0/1
127.0.0.0/8         Direct 0    0     D     127.0.0.1   InLoopBack0
127.0.0.1/32        Direct 0    0     D     127.0.0.1   InLoopBack0
127.255.255.255/32  Direct 0    0     D     127.0.0.1   InLoopBack0
255.255.255.255/32  Direct 0    0     D     127.0.0.1   InLoopBack0
```

（4）在 R1 上测试到 R3 的连通性，代码如下。

```
[R1]ping 23.1.1.3
 PING 23.1.1.3: 56  data bytes, press CTRL_C to break
   Reply from 23.1.1.3: bytes=56 Sequence=1 ttl=254 time=20 ms
   Reply from 23.1.1.3: bytes=56 Sequence=2 ttl=254 time=30 ms
   Reply from 23.1.1.3: bytes=56 Sequence=3 ttl=254 time=30 ms
   Reply from 23.1.1.3: bytes=56 Sequence=4 ttl=254 time=40 ms
   Reply from 23.1.1.3: bytes=56 Sequence=5 ttl=254 time=30 ms

--- 23.1.1.3 ping statistics ---
   5 packet(s) transmitted
   5 packet(s) received
   0.00% packet loss
   round-trip min/avg/max = 20/30/40 ms
```

3. 等价路由

所谓等价路由就是路由器通过路由选举规则无法比较优劣的路由。等价路由拓扑示意如图 3-14 所示。当网络设备有多条去往目的网段的路径，并且这些路径的带宽一致或带宽差异不大时，可以使用等价路由。

V3-2　配制静态路由（等价路由）

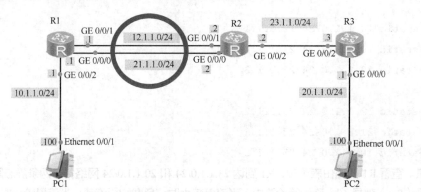

图 3-14　等价路由拓扑示意

（1）配置端口 IP 地址，代码如下。

```
[R1]int g0/0/0
[R1-GigabitEthernet0/0/0]ip add 21.1.1.1 24
[R1-GigabitEthernet0/0/0]int g0/0/1
[R1-GigabitEthernet0/0/1]ip add 12.1.1.1 24
[R1-GigabitEthernet0/0/1]int g0/0/2
[R1-GigabitEthernet0/0/2]ip add 10.1.1.1 24
[R1-GigabitEthernet0/0/2]

[R2]int g0/0/0
[R2-GigabitEthernet0/0/0]ip add 21.1.1.2 24
[R2-GigabitEthernet0/0/0]int g0/0/1
[R2-GigabitEthernet0/0/1]ip add 12.1.1.2 24
[R2-GigabitEthernet0/0/1]int g0/0/2
[R2-GigabitEthernet0/0/2]ip add 23.1.1.2 24
[R2-GigabitEthernet0/0/2]

[R3]int g0/0/0
[R3-GigabitEthernet0/0/0]ip add 20.1.1.1 24
[R3-GigabitEthernet0/0/0]int g0/0/2
[R3-GigabitEthernet0/0/2]ip add 23.1.1.3 24
[R3-GigabitEthernet0/0/2]
```

（2）配置静态路由。在 R1 上手动配置到 23.1.1.0/24 和 20.1.1.0/24 网络的所有静态路由；在 R2 上手动配置到 10.1.1.0/24 和 20.1.1.0/24 网络的所有静态路由；在 R3 上手动配置到 21.1.1.0/24、12.1.1.0/24 和 10.1.1.0/24 网络的所有静态路由，具体代码如下。

```
[R1]ip route-static 23.1.1.0 24 12.1.1.2
[R1]ip route-static 23.1.1.0 24 21.1.1.2
[R1]ip route-static 20.1.1.0 24 12.1.1.2
[R1]ip route-static 20.1.1.0 24 21.1.1.2

[R2]ip route-static 10.1.1.0 24 12.1.1.1
[R2]ip route-static 10.1.1.0 24 21.1.1.1
[R2]ip route-static 20.1.1.0 24 23.1.1.3

[R3]ip route-static 12.1.1.0 24 23.1.1.2
[R3]ip route-static 21.1.1.0 24 23.1.1.2
[R3]ip route-static 10.1.1.0 24 23.1.1.2
```

（3）查看路由表。查看 R1 的路由表可知，R1 到达 23.1.1.0/24 和 20.1.1.0/24 网络各有两条静态路由。静态路由优先级（Pre）都是 60，称为等价路由。当流量较大时，负载均衡流量；当流量较小时，选择其中的一条路由。具体代码如下。

```
[R1]display ip routing-table
[R1]dis ip routing-table
Route Flags: R - relay, D - download to fib
------------------------------------------------------------------------------
Routing Tables: Public
         Destinations : 15      Routes : 17
Destination/Mask    Proto   Pre  Cost  Flags  NextHop        Interface
10.1.1.0/24         Direct  0    0     D      10.1.1.1       GigabitEthernet0/0/2
10.1.1.1/32         Direct  0    0     D      127.0.0.1      GigabitEthernet0/0/2
10.1.1.255/32       Direct  0    0     D      127.0.0.1      GigabitEthernet0/0/2
12.1.1.0/24         Direct  0    0     D      12.1.1.1       GigabitEthernet0/0/1
12.1.1.1/32         Direct  0    0     D      127.0.0.1      GigabitEthernet0/0/1
12.1.1.255/32       Direct  0    0     D      127.0.0.1      GigabitEthernet0/0/1
20.1.1.0/24         Static  60   0     RD     12.1.1.2       GigabitEthernet0/0/1
                    Static  60   0     RD     21.1.1.2       GigabitEthernet0/0/0
21.1.1.0/24         Direct  0    0     D      21.1.1.1       GigabitEthernet0/0/0
21.1.1.1/32         Direct  0    0     D      127.0.0.1      GigabitEthernet0/0/0
21.1.1.255/32       Direct  0    0     D      127.0.0.1      GigabitEthernet0/0/0
23.1.1.0/24         Static  60   0     RD     12.1.1.2       GigabitEthernet0/0/1
                    Static  60   0     RD     21.1.1.2       GigabitEthernet0/0/0
127.0.0.0/8         Direct  0    0     D      127.0.0.1      InLoopBack0
```

（4）路由测试。由于流量较小，测试看不出负载均衡，具体代码如下。

```
//在 PC1 上测试 PC2
PC1>tracert 20.1.1.100
traceroute to 20.1.1.100, 8 hops max
(ICMP), press Ctrl+C to stop
 1  10.1.1.1    15 ms  16 ms  <1 ms
 2  12.1.1.2    47 ms  16 ms  <1 ms
 3  23.1.1.3    46 ms  32 ms  15 ms
 4  *20.1.1.100    16 ms  16 ms
//在 PC2 上测试 PC1
PC2>tracert 10.1.1.100
traceroute to 10.1.1.100, 8 hops max
(ICMP), press Ctrl+C to stop
 1  20.1.1.1    16 ms  15 ms  16 ms
 2  23.1.1.2    16 ms  15 ms  16 ms
 3  12.1.1.1    16 ms  15 ms  31 ms
 4  10.1.1.100    16 ms  16 ms  31 ms
```

4. 浮动路由

所谓浮动路由，即当多条链路带宽差异特别大时，使高带宽链路成为主链路，低带宽链路成为备份链路。根据路由的选举规则调整路由的优劣，让高带宽链路更优。可以将高带宽链路的优先级调低，或将低带宽链路的优先级调高。路由的选举规则如下。

① 比较子网掩码长度，越长越优，子网掩码越长，表示网段越精确。
② 比较路由优先级，不同的路由协议拥有不同的优先级，优先级越小越好。
③ 比较路由开销值，开销值描述了从路由器本地到达目的网段的距离有多远，越小越优。

浮动路由拓扑示意如图 3-15 所示，Serial 链路的传输速率远低于 GE 链路的传输速率，所以可以把 GE 链路设置为主链路，Serial 链路设置为备份链路。

V3-3 配置静态路由（浮动路由）

图 3-15 浮动路由拓扑示意

（1）配置端口 IP 地址，代码如下。

```
[R1]int g0/0/0
[R1-GigabitEthernet0/0/0]ip add 21.1.1.1 24
[R1-GigabitEthernet0/0/0]int Serial 1/0/0
[R1-Serial1/0/0]ip add 12.1.1.1 24
[R1-GigabitEthernet0/0/1]int g0/0/2
[R1-GigabitEthernet0/0/2]ip add 10.1.1.1 24
[R1-GigabitEthernet0/0/2]

[R2]int g0/0/0
[R2-GigabitEthernet0/0/0]ip add 21.1.1.2 24
[R2-GigabitEthernet0/0/0]int Serial 1/0/0
[R2-Serial1/0/0]ip add 12.1.1.2 24
[R2-GigabitEthernet0/0/1]int g0/0/2
[R2-GigabitEthernet0/0/2]ip add 23.1.1.2 24
[R2-GigabitEthernet0/0/2]


[R3]int g0/0/0
```

```
[R3-GigabitEthernet0/0/0]ip add 20.1.1.1 24
[R3-GigabitEthernet0/0/0]int g0/0/2
[R3-GigabitEthernet0/0/2]ip add 23.1.1.3 24
[R3-GigabitEthernet0/0/2]
```

（2）配置浮动路由。静态路由优先级为 60，把备份路由优先级修改为比 60 大即可实现备份路由。这里将 Serial 链路的静态路由的 Preference 参数修改为 100，则优先级为 100 的那条路由为备份路由，将不会在路由表中显示，具体代码如下。

```
[R1]ip route-static 23.1.1.0 24 12.1.1.2 preference 100
[R1]ip route-static 23.1.1.0 24 21.1.1.2
[R1]ip route-static 20.1.1.0 24 12.1.1.2 preference 100
[R1]ip route-static 20.1.1.0 24 21.1.1.2


[R2]ip route-static 10.1.1.0 24 12.1.1.1 preference 100
[R2]ip route-static 10.1.1.0 24 21.1.1.1
[R2]ip route-static 20.1.1.0 24 23.1.1.3


[R3]ip route-static 12.1.1.0 24 23.1.1.2
[R3]ip route-static 21.1.1.0 24 23.1.1.2
[R3]ip route-static 10.1.1.0 24 23.1.1.2
```

（3）查看路由表。查看 R1 的路由表可知，上面配置的优先级为 100 的备份路由（即 Serial 链路的静态路由）没有显示在路由表中。在路由表中只看到优先级为 60 的静态路由，即 GE 链路的静态路由。当流量经过 R1、R2 区域的时候，数据会通过 21.1.1.0/24 网络的 GE 链路。当 GE 链路出现故障时，路由表立即激活优先级为 100 的 Serial 链路的静态路由，从而保证网络的可靠性，具体代码如下。

```
[R1]dis ip routing-table
Route Flags: R - relay, D - download to fib
------------------------------------------------------------------------
Routing Tables: Public
        Destinations : 16      Routes : 16
Destination/Mask    Proto    Pre  Cost  Flags   NextHop         Interface
10.1.1.0/24         Direct   0    0     D       10.1.1.1        GigabitEthernet0/0/2
10.1.1.1/32         Direct   0    0     D       127.0.0.1       GigabitEthernet0/0/2
10.1.1.255/32       Direct   0    0     D       127.0.0.1       GigabitEthernet0/0/2
12.1.1.0/24         Direct   0    0     D       12.1.1.1        Serial1/0/0
12.1.1.1/32         Direct   0    0     D       127.0.0.1       Serial1/0/0
12.1.1.2/32         Direct   0    0     D       12.1.1.2        Serial1/0/0
12.1.1.255/32       Direct   0    0     D       127.0.0.1       Serial1/0/0
20.1.1.0/24         Static   60   0     RD      21.1.1.2        GigabitEthernet0/0/0
21.1.1.0/24         Direct   0    0     D       21.1.1.1        GigabitEthernet0/0/0
21.1.1.1/32         Direct   0    0     D       127.0.0.1       GigabitEthernet0/0/0
21.1.1.255/32       Direct   0    0     D       127.0.0.1       GigabitEthernet0/0/0
```

```
23.1.1.0/24              Static   60  0  RD  21.1.1.2    GigabitEthernet0/0/0
127.0.0.0/8              Direct   0   0  D   127.0.0.1   InLoopBack0
127.0.0.1/32             Direct   0   0  D   127.0.0.1   InLoopBack0
127.255.255.255/32       Direct   0   0  D   127.0.0.1   InLoopBack0
255.255.255.255/32       Direct   0   0  D   127.0.0.1   InLoopBack0
```

（4）路由测试。如图 3-15 所示，当 PC1 与 PC2 通信经过 R1、R2 区域时，数据会通过 GE 链路网络 21.1.1.0/24，不会通过 Serial 链路网络 12.1.1.0/24，具体代码如下。

```
//在 PC1 上测试 PC2
PC>tracert 20.1.1.100
traceroute to 20.1.1.100, 8 hops max
(ICMP), press Ctrl+C to stop
 1  10.1.1.1      16 ms  16 ms  15 ms
 2  *21.1.1.2     31 ms  16 ms
 3  *23.1.1.3     31 ms  16 ms
 4  *20.1.1.100   16 ms  15 ms


//在 PC2 上测试 PC1
PC>tracert 10.1.1.100
traceroute to 10.1.1.100, 8 hops max
(ICMP), press Ctrl+C to stop
 1  20.1.1.1      16 ms  15 ms  16 ms
 2  23.1.1.2      16 ms  15 ms  16 ms
 3  21.1.1.1      16 ms  15 ms  31 ms
 4  10.1.1.100    16 ms  16 ms  31 ms
```

当关闭 R1 的 GE 0/0/0 端口时，PC1 与 PC2 通信经过 R1、R2 区域时，因为 GE 链路网络 21.1.1.0/24 中断，所以数据立即通过 Serial 链路网络 12.1.1.0/24，具体代码如下。

```
//关闭 R1 的 GE 0/0/0 端口
[R1]int g0/0/0
[R1-GigabitEthernet0/0/0]shutdown
//在 PC1 上再次测试 PC2
PC>tracert 20.1.1.100
traceroute to 20.1.1.100, 8 hops max
(ICMP), press Ctrl+C to stop
 1  10.1.1.1    <1 ms  15 ms  16 ms
 2  12.1.1.2    15 ms  <1 ms  16 ms


//在 PC2 上再次测试 PC1
PC>tracert 10.1.1.100
traceroute to 10.1.1.100, 8 hops max
(ICMP), press Ctrl+C to stop
```

```
 1  20.1.1.1      15 ms   16 ms   <1 ms
 2  23.1.1.2      31 ms   16 ms   15 ms
 3  12.1.1.1      32 ms   15 ms   16 ms
 4  10.1.1.100    31 ms   16 ms   31 ms
```

（5）再次查看 R1 的路由表。GE 链路的静态路由信息更新已经中断，在路由表中没有数据，优先级为 100 的 Serial 链路的静态路由信息出现在路由表中，具体代码如下。

```
[R1]dis ip routing-table
Route Flags: R - relay, D - download to fib
------------------------------------------------------------------------------
Routing Tables: Public
         Destinations : 13       Routes : 13
Destination/Mask        Proto   Pre Cost    Flags   NextHop         Interface
10.1.1.0/24             Direct  0   0       D       10.1.1.1        GigabitEthernet0/0/2
10.1.1.1/32             Direct  0   0       D       127.0.0.1       GigabitEthernet0/0/2
10.1.1.255/32           Direct  0   0       D       127.0.0.1       GigabitEthernet0/0/2
12.1.1.0/24             Direct  0   0       D       12.1.1.1        Serial1/0/0
12.1.1.1/32             Direct  0   0       D       127.0.0.1       Serial1/0/0
12.1.1.2/32             Direct  0   0       D       12.1.1.2        Serial1/0/0
12.1.1.255/32           Direct  0   0       D       127.0.0.1       Serial1/0/0
20.1.1.0/24             Static  100 0       RD      12.1.1.2        Serial1/0/0
23.1.1.0/24             Static  100 0       RD      12.1.1.2        Serial1/0/0
127.0.0.0/8             Direct  0   0       D       127.0.0.1       InLoopBack0
127.0.0.1/32            Direct  0   0       D       127.0.0.1       InLoopBack0
127.255.255.255/32      Direct  0   0       D       127.0.0.1       InLoopBack0
255.255.255.255/32      Direct  0   0       D       127.0.0.1       InLoopBack0
```

5. 默认路由

所谓默认路由，是指在路由信息中前缀全 0 且子网掩码全 0 的路由，即 0.0.0.0/0，这种路由可以代表所有网段。一般在园区网的出口路由器上配置默认路由，以实现对公网内所有网段的访问。默认路由拓扑示意如图 3-16 所示。假如 R2 是外网路由器，R1 是内网路由器。

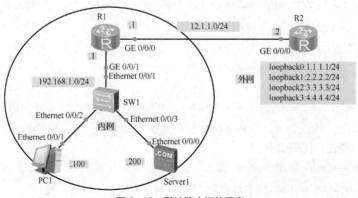

图 3-16　默认路由拓扑示意

（1）配置端口 IP 地址，代码如下。

```
[R1]int g0/0/0
[R1-GigabitEthernet0/0/0]ip add 12.1.1.1 24
Error: The specified IP address is invalid.
[R1-GigabitEthernet0/0/0]int g0/0/1
[R1-GigabitEthernet0/0/1]ip add 192.168.1.1 24
[R1-GigabitEthernet0/0/1]

[R2]int g0/0/0
[R2-GigabitEthernet0/0/0]ip add 12.1.1.2 24
[R2-GigabitEthernet0/0/0]int loopback 0
[R2-LoopBack0]ip add 1.1.1.1 24
[R2-LoopBack0]int loopback 1
[R2-LoopBack1]ip add 2.2.2.2 24
[R2-LoopBack1]int loopback 2
[R2-LoopBack2]ip add 3.3.3.3 24
[R2-LoopBack2]int loopback 3
[R2-LoopBack3]ip add 4.4.4.4 24
[R2]dis ip int brief
*down: administratively down
^down: standby
(l): loopback
(s): spoofing
The number of interface that is UP in Physical is 6
The number of interface that is DOWN in Physical is 2
The number of interface that is UP in Protocol is 6
The number of interface that is DOWN in Protocol is 2
Interface                IP Address/Mask    Physical    Protocol
GigabitEthernet0/0/0     12.1.1.2/24        up          up
GigabitEthernet0/0/1     unassigned         down        down
GigabitEthernet0/0/2     unassigned         down        down
LoopBack0                1.1.1.1/24         up          up(s)
LoopBack1                2.2.2.2/24         up          up(s)
LoopBack2                3.3.3.3/24         up          up(s)
LoopBack3                4.4.4.4/24         up          up(s)
NULL0                    unassigned         up          up(s)
```

（2）配置默认路由，代码如下。

```
[R1]ip route-static 0.0.0.0 0 12.1.1.2
[R2]ip route-static 192.168.1.0 24 12.1.1.1
```

（3）查看路由表，代码如下。

```
[R1]dis ip routing-table
Route Flags: R - relay, D - download to fib
------------------------------------------------------------------------
Routing Tables: Public
         Destinations : 11       Routes : 11
Destination/Mask    Proto   Pre  Cost  Flags  NextHop        Interface
0.0.0.0/0           Static  60   0     RD     12.1.1.2       GigabitEthernet0/0/0
12.1.1.0/24         Direct  0    0     D      12.1.1.1       GigabitEthernet0/0/0
12.1.1.1/32         Direct  0    0     D      127.0.0.1      GigabitEthernet0/0/0
12.1.1.255/32       Direct  0    0     D      127.0.0.1      GigabitEthernet0/0/0
127.0.0.0/8         Direct  0    0     D      127.0.0.1      InLoopBack0
127.0.0.1/32        Direct  0    0     D      127.0.0.1      InLoopBack0
127.255.255.255/32  Direct  0    0     D      127.0.0.1      InLoopBack0
192.168.1.0/24      Direct  0    0     D      192.168.1.1    GigabitEthernet0/0/1
192.168.1.1/32      Direct  0    0     D      127.0.0.1      GigabitEthernet0/0/1
192.168.1.255/32    Direct  0    0     D      127.0.0.1      GigabitEthernet0/0/1
255.255.255.255/32  Direct  0    0     D      127.0.0.1      InLoopBack0
```

（4）网络测试。在内网 PC1 上访问外网 1.1.1.0/24 网络中的主机地址 1.1.1.1/24，代码如下。

```
PC>tracert 1.1.1.1
traceroute to 1.1.1.1, 8 hops max
(ICMP), press Ctrl+C to stop
 1 192.168.1.1   31 ms   47 ms   47 ms
 2 1.1.1.1   47 ms   31 ms   47 ms

PC>ping 1.1.1.1
Ping 1.1.1.1: 32 data bytes, Press Ctrl_C to break
Request timeout!
From 1.1.1.1: bytes=32 seq=2 ttl=254 time=62 ms
From 1.1.1.1: bytes=32 seq=3 ttl=254 time=32 ms
From 1.1.1.1: bytes=32 seq=4 ttl=254 time=16 ms
From 1.1.1.1: bytes=32 seq=5 ttl=254 time=47 ms
--- 1.1.1.1 ping statistics ---
  5 packet(s) transmitted
  4 packet(s) received
  20.00% packet loss
  round-trip min/avg/max = 0/39/62 ms
```

【任务拓展】

静态路由拓扑示意如图 3-17 所示，厘清每个端口的 IP 地址和各网段的 IP 地址，配置各 PC 和各路由端口的 IP 地址，各直连设备 ping 通之后，实现静态路由配置，使 3 台 PC 间能相互进行通信。

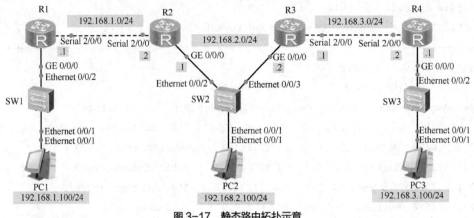

图 3-17 静态路由拓扑示意

任务 3.2 动态路由的配置

【知识准备】

3.2.1 动态路由协议

动态路由是指路由器使用路由协议（Routing Protocol）获悉网络，并更新路由表。路由器能够根据路由器之间交换的特定路由信息自动建立自己的路由表，并根据链路和节点的变化适时地进行自动调整。当网络中节点或节点间的链路发生故障，或存在其他可用路由时，动态路由可以自行选择最佳的可用路由并继续转发报文。

动态路由协议是路由器之间交互信息的一种语言，定义了一套路由器之间通信时使用的规则，以维护路由表、提供最佳转发路径。路由器之间通过路由协议共享网络状态和网络可达性的一些信息。使用路由协议从其他路由器那里获悉路由，当网络拓扑发生变化时，路由器会自动更新路由信息。相较于静态路由，动态路由的优点在于不需要手动书写路由，在配置好端口 IP 地址和直连网段后，路由表之间可以相互学习。但这并不代表动态路由可以取代静态路由，在网络中静态路由和动态路由相互补充。

常见的动态路由协议的类别示意如图 3-18 所示。

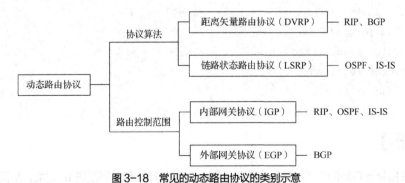

图 3-18 常见的动态路由协议的类别示意

基于协议算法不同，可将动态路由协议分成距离矢量路由协议（Distance Vector Routing Protocol，DVRP）和链路状态路由协议（Link State Routing Protocol，LSRP）两类。

1. DVRP

在距离矢量算法中，每台路由器只有与它直接相连的路由器的信息，没有网络中每台路由器的信息。DVRP 指的是基于距离矢量算法的路由协议，包括边界网关协议（Border Gateway Protocol，BGP）和路由信息协议（Routing Information Protocol，RIP）。其中，BGP 也被称为路径矢量（Path Vector）路由协议；RIP 是最具代表性的 DVRP。"距离矢量"这个概念包含"距离"和"方向"两个关键的信息，其中"距离"指到达目的网络的度量值，而"方向"指到达该目的网络的下一跳设备。

每一台运行 DVRP 的路由器都会周期性地将自己的路由表通告出去，其直连的路由器会收到这些路由信息，在学习前者通告的路由并更新自己的路由表后，它也会向自己直连的路由器通告其路由表，最终网络中的每台路由器都能获知到达各个网段的路由，这个过程被称为路由的泛洪（Flooding）过程。

DVRP 工作机制如图 3-19 所示。

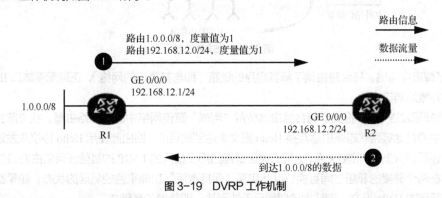

图 3-19　DVRP 工作机制

在图 3-19 中，R1 和 R2 两台路由器直连。初始情况下 R1 和 R2 都只知道自己"家门口的情况"，也就是说，R1 和 R2 都自动发现了自己直连端口的路由。R1 在其路由表中写入 192.168.12.0/24 和 1.0.0.0/8 两条直连路由，而 R2 则在其路由表中写入 192.168.12.0/24 这一条直连路由。当然，此刻 R2 是无法访问 1.0.0.0/8 的，因为在它的路由表中并没有任何能够到达这个网段的路由信息。

在 R1 和 R2 上使用 DVRP 后，R1 将已经发现的路由（1.0.0.0/8 和 192.168.12.0/24）通过路由协议报文通告给 R2，这两条路由各附带一个度量值。以 DVRP 的典型代表 RIP 为例，它使用跳数（Hop Count）作为路由的度量值。所谓跳数，就是到达目的网段所需经过的路由器的台数，直连网段的度量值被视为 0 跳。R1 将两条直连路由通告给 R2 时，为路由设置的跳数为 1。因为，"自家门口"的这些网段对于自己而言可以直接到达（只需 0 跳），现在"别人"要经过"我家"来访问该网段，就需要加上"我这一跳"。

由于 192.168.12.0/24 是 R2 自己的直连网段，因此 R2 会忽略 R1 通告过来的到达该网段的路由更新。而 1.0.0.0/8 路由对于 R2 而言是未知的，因此 R2 将这条路由加载进路由表中，同时为该条路由关联度量值为 1 跳，并且把路由的通告者 R1 视为该路由的下一跳。于是当 R2 要转发到达 1.0.0.0/8 的数据包时，就会将数据包发给下一跳路由器 R1，即"R2 认为 1.0.0.0/8 可以通过 R1 到达，自己与该网段距离 1 跳路由器"，这就是名称"距离矢量"的由来。运行 DVRP 的路由器并不了解网络的拓扑结构，该路由器只知道自己与目的网络之间的距离，以及从哪个方向可到达目的网络。

2. LSRP

LSRP 指的是基于链路状态算法的路由协议，包括开放最短路径优先（Open Shortest Path First，OSPF）协议和中间系统到中间系统（Intermediate System to Intermediate System，IS-IS）协议，其又称

为最短路径优先路由协议，每台路由器都拥有网络中其他路由器的全部信息，以及网络的流量状态。

运行 LSRP 的路由器，只将它所直连的链路的状态与邻居共享，邻居是指一个域（Domain）内或一个区域（Area）内的所有路由器。LSRP 过程如图 3-20 所示，详细工作过程如下。

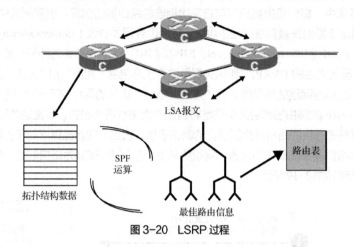

图 3-20　LSRP 过程

① 了解直连网络。每台路由器了解其自身的链路（即与其直连的网络），正确配置端口 IP 地址和子网掩码并激活端口。

② 向邻居发送 Hello 报文。每台路由器负责"问候"直连网络中的相邻路由器。状态路由器通过直连网络中的其他链路状态路由器互换 Hello 报文来达到此目的。路由器使用 Hello 协议来发现其链路上的所有邻居，形成一种邻接关系，这里的邻居是指启用了相同的 LSRP 的其他任何路由器。这些 Hello 报文持续在两个邻接的邻居之间互换，以此实现"保持激活"功能来监控邻居的状态。如果路由器不再收到某邻居的 Hello 报文，则认为该邻居已无法到达，即邻接关系破裂。

③ 创建链路状态数据包（Link State Packet，LSP）。每台路由器创建一个 LSP，其中包含与该路由器直连的每条链路的信息，如链路类型、IP 地址、子网掩码等。一旦建立了邻接关系，即可创建 LSP，并仅向建立邻接关系的路由器发送 LSP。

④ 将 LSP 泛洪给邻居。每台路由器将 LSP 泛洪给所有邻居，然后邻居将收到的所有 LSP 存储到数据库中。各个邻居将 LSP 泛洪给自己的邻居，直到区域内的所有路由器均收到那些 LSP 为止。每台路由器会在本地数据库中存储邻居发来的 LSP 的副本。路由器将其链路状态信息泛洪到路由区域内的其他所有路由器，它一旦收到来自邻居的 LSP，不经过中间计算，就立即将这个 LSP 发给所有端口，此过程在整个路由区域内的所有路由器上形成 LSP 的泛洪效应。

⑤ 构建链路状态数据库（Link State Database，LSDB）。经过 LSP 泛洪过程后，每台路由器都将拥有来自整个路由区域内所有路由器的 LSP，并利用该信息构建完整的链路状态数据库。链路状态数据库构建成功后，各个路由器即可利用该数据库中存储的信息和最短路径优先（Shortest Path First，SPF）算法计算通向每个网络的首选路径（即最短路径）。

基于路由控制范围的不同，可将动态路由协议分成内部网关协议（Interior Gateway Protocol，IGP）和外部网关协议（Exterior Gateway Protocol，EGP）。

1. 内部网关协议

内部网关协议是指在一个自治系统（Autonomous System，AS）内部所使用的一种路由协议。IGP 适用于单个 ISP 的统一路由协议的运行，一般由一个 ISP 运营的网络位于一个自治系统内，有统一的自治系统号。

2. 外部网关协议

BGP 是自治系统间的路由协议，是一种外部网关协议，多用于 Internet，在不同运营商之间交换路由信息，在某些大型的企业的内部网络中有时也会用到 BGP 路由协议。

路由协议作用范围示意如图 3-21 所示。

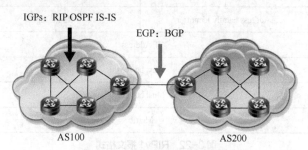

图 3-21　路由协议作用范围示意

本书着重讲解 RIP 和 OSPF 协议，若读者想了解其他路由协议，则可以查阅相关资料。

3.2.2　RIP 概述

1. RIP 介绍

RIP 是基于距离矢量算法的协议，它使用跳数即 Metric 来衡量到达目的地址的路由距离，只与自己相邻的路由器交换信息，范围限制在 15 跳之内。因此 RIP 主要应用于规模较小的网络中，适用于大多数的校园网和使用速率变化不是很大的区域性网络。对于更为复杂的环境和大型网络，不使用 RIP。RIP 存在收敛慢、易产生路由环路、可扩展性差的问题，目前已逐渐被 OSPF 协议取代。

默认情况下，在 RIP 网络中，设备与它直接相连的网络的跳数为 0，通过一台设备可达的网络的跳数为 1，其余以此类推。也就是说，跳数等于从本网络到达目的网络间的设备数量。为限制收敛时间，RIP 规定跳数取 0~15 之间的整数，大于或等于 16 的跳数被定义为无穷大，即目的网络或主机不可达。这个限制使 RIP 不可能在大型网络中得到应用。

RIP 具有以下特点。

（1）每台路由器只和相邻的路由器交换信息。

（2）路由器交换的信息指的是当前路由器所知道的全部信息，即路由表，RIP 的路由条目是到某个网络的最短路径，以及下一跳 IP 地址。

（3）路由器按固定的时间交换信息，如每隔 30s 交换一次信息。

2. RIP 报文

RIP 主要有 RIPv1 和 RIPv2 两个版本。

RIPv1 是有类别路由协议（Classful Routing Protocol），只支持以广播方式发送报文。RIPv1 的报文中没有携带子网掩码信息，只能识别 A、B、C 类这样的自然网段的路由，因此 RIPv1 无法支持路由聚合，也不支持不连续子网（Discontinuous Subnet）。

RIPv2 是无类别路由协议（Classless Routing Protocol），其有两种报文发送方式，即广播方式和组播方式，默认采用组播方式发送报文，使用的组播地址为 224.0.0.9。当端口运行 RIPv2 广播方式时，也可接收 RIPv1 的报文。

（1）RIPv1 报文格式

RIP 报文由头部（Header）和多个路由条目（Route Entries）部分组成。在一个 RIP 报文中，最多可以有 25 个路由条目。RIP 基于 UDP，并且 RIPv1 报文不能超过 512 字节。RIPv1 报文格式如图 3-22 所示，各字段含义如表 3-3 所示。

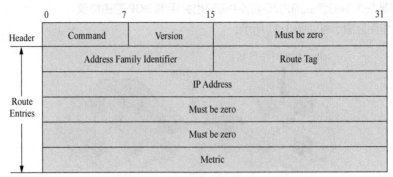

图 3-22　RIPv1 报文格式

表 3-3　RIPv1 报文各字段含义

字段名	长度/位	含义
Command（命令）	8	标识报文的类型： 1 表示 Request 报文，向邻居请求全部或部分路由信息； 2 表示 Response 报文，发送自己全部或部分路由信息，一个 Response 报文中最多包含 25 个路由条目
Version（版本）	8	RIP 的版本号：1 表示 RIPv1，2 表示 RIPv2
Must be zero	16	这是一个保留字段，用 0 填充，只出现在报文头
Address Family Identifier（AFI，地址族标识）	16	其值为 2 时表示 IP。对于 Request 报文，此字段值为 0
Route Tag（路由标记）	16	标记外部路由或者引入到 RIPv2 协议中的路由
IP Address（IP 地址）	32	该路由的目的 IP 地址，可以是自然网段的地址，也可以是子网地址或主机地址
Metric（度量值）	32	路由的度量值。对于 Request 报文，此字段值为 16

（2）RIPv2 报文格式

RIPv2 报文格式如图 3-23 所示，各字段含义如表 3-4 所示。RIPv2 为了支持报文验证，使用第一个路由条目作为验证项，并将 AFI 字段的值设为 0xFFFF 作为标识。

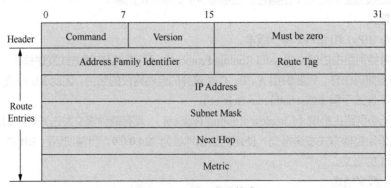

图 3-23　RIPv2 报文格式

表 3-4　RIPv2 报文各字段含义

字段名	长度/位	含义
Command	8	标识报文的类型： 1 表示 Request 报文，向邻居请求全部或部分路由信息； 2 表示 Response 报文，发送自己全部或部分路由信息，一个 Response 报文中最多包含 25 个路由条目
Version	8	RIP 的版本号：1 表示 RIPv1，2 表示 RIPv2
Must be zero	16	必须为零字段
Address Family Identifier	16	地址族标识，其值为 2 时表示 IP。对于 Request 报文，此字段值为 0
Route Tag（外部路由标记）	16	外部路由标记
IP Address	32	该路由的目的 IP 地址，可以是自然网段的地址，也可以是子网地址或主机地址
Subnet Mask（子网掩码）	32	目的地址的子网掩码
Next Hop（下一跳地址）	32	提供一个更好的下一跳 IP 地址。如果为 0.0.0.0，则表示发布此路由的路由器地址就是最优下一跳 IP 地址
Metric	32	路由的度量值。对于 Request 报文，此字段值为 16

RIPv2 验证报文格式如图 3-24 所示，各字段含义如表 3-5 所示。

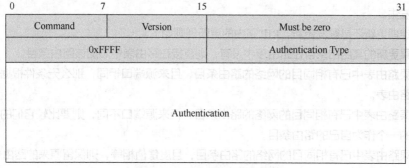

图 3-24　RIPv2 验证报文格式

表 3-5　RIPv2 验证报文各字段含义

字段名	长度/位	含义
Command	8	标识报文的类型： 1 表示 Request 报文，向邻居请求全部或部分路由信息； 2 表示 Response 报文，发送自己全部或部分路由信息，一个 Response 报文中最多包含 25 个路由条目
Version	8	RIP 的版本号：1 表示 RIPv1，2 表示 RIPv2
Must be zero	16	必须为零字段
0xFFFF（验证项标识）	16	表示整个路由报文需要验证
Authentication Type（验证类型）	16	1 表示明文验证，2 表示 MD5 验证
Authentication（身份验证）	16	当使用明文验证时，该字段才会包含密码信息

（3）RIPv1 和 RIPv2 的区别

RIPv1 和 RIPv2 的区别如表 3-6 所示。

表 3-6 RIPv1 和 RIPv2 的区别

RIPv1	RIPv2
有类别路由协议	无类别路由协议
不支持可变长子网掩码（Variable Length Subnet Mask，VLSM）	支持 VLSM
没有验证功能	支持验证功能，并且有明文和 MD5 两种验证方式
没有手动汇总的功能	可以在关闭自动汇总的前提下进行手动汇总
广播更新	组播更新
不能对路由进行标记	可以对路由进行标记，用于过滤和制定策略
发送的 update 包最多可以携带 25 个路由条目	在有验证的情况下最多只能携带 24 个路由条目
发送的 update 包中没有 next-hop 属性	有 next-hop 属性，可以用于路由更新的重定向

3. RIP 路由更新

RIP 有定期更新和触发更新两种更新机制。定期更新是根据设置的更新计时器定期发送 RIP 路由通告；触发更新是 RIP 路由器一旦察觉到网络变化，就尽快甚至是立即发送更新报文，而不等待更新周期结束。只要触发更新的速度足够快，就可以大大防止"计数到无穷大"的现象发生，但是这一现象仍有可能发生。

无论是定期更新还是触发更新，RIP 路由的更新规则如下。

（1）如果更新的某路由条目在路由表中没有，则直接在路由表中添加该路由条目。

（2）如果路由表中已有相同目的网络的路由条目，且来源端口相同，那么无条件根据最新的路由信息更新其路由表。

（3）如果路由表中已有相同目的网络的路由条目，但来源端口不同，则要比较它们的度量值，将度量值较小的一个作为自己的路由条目。

（4）如果路由表中已有相同目的网络的路由条目，且度量值相等，则保留原来的路由条目。

3.2.3 RIP 原理

RIP 路由拓扑示意如图 3-25 所示。

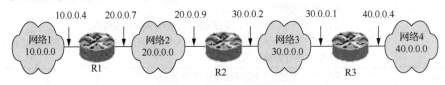

图 3-25 RIP 路由拓扑示意

1. RIP 初始路由

初始情况下，所有路由器中的路由表中都只有自己所直接连接的网络的路由条目信息，不是 RIP 路由条目，而是直连路由条目，无须下一跳（用"--"表示），度量值（用"距离"表示）也均为 0。各路由器的初始路由表如图 3-26 所示，均只有两个直连网络的路由条目。

R1的路由表			R2的路由表			R3的路由表		
目的网络地址	下一跳IP地址	距离	目的网络地址	下一跳IP地址	距离	目的网络地址	下一跳IP地址	距离
10.0.0.0	--	0	20.0.0.0	--	0	30.0.0.0	--	0
20.0.0.0	--	0	30.0.0.0	--	0	40.0.0.0	--	0

图 3-26　各路由器的初始路由表

2. R2 的路由表更新

各路由器会按设置的周期（默认为 30s）向邻居路由器发送路由更新。具体哪台路由器先发送路由更新，取决于哪台路由器先启动。现假设路由器 R2 先收到来自路由器 R1 和 R3 的路由更新，然后更新了自己的路由表，如图 3-27 所示。从中可以看出，它新添加了分别通过 R1 和 R3 到达 10.0.0.0 网络和 40.0.0.0 网络的路由条目，度量值均为 1，因为它只经过了 1 跳。

3. R1 从 R2 获取路由信息

R2 更新自己的路由表后，会把完整的路由表发给邻居路由器 R1 和 R3。R1 和 R3 分别进行更新。根据前面介绍的 RIP 路由更新的规则可以知道，R1 对每个度量值加 1，得到的路由表如图 3-28 所示。

目的网络地址	下一跳IP地址	距离
20.0.0.0	--	0
30.0.0.0	--	0
10.0.0.0	20.0.0.7	1
40.0.0.0	30.0.0.1	1

图 3-27　R2 的路由表更新

目的网络地址	下一跳IP地址	距离
20.0.0.0	20.0.0.9	1
30.0.0.0	20.0.0.9	1
10.0.0.0	20.0.0.9	2
40.0.0.0	20.0.0.9	2

图 3-28　R1 从 R2 获取路由信息

4. R1 的路由表更新

R1 再把图 3-28 所示的路由表与自己原来的路由表进行比较，凡是新添加的和度量值小于或等于原来的路由条目均将更新，度量值更大的路由条目将忽略更新。经过比较发现有两个新的路由条目，其目的网络分别为 30.0.0.0 和 40.0.0.0，直接在路由表中添加这两个条目。而原来已有的两个条目 10.0.0.0 和 20.0.0.0，发现现在的度量值 1 比原来的 0 大，忽略更新，即可得到 R1 更新后的路由表，如图 3-29 所示。

目的网络地址	下一跳IP地址	距离
10.0.0.0	--	0
20.0.0.0	--	0
30.0.0.0	20.0.0.9	1
40.0.0.0	20.0.0.9	2

图 3-29　R1 的路由表更新

3.2.4　RIP 的防环机制

对于具备路由功能的网络设备来说，当其执行数据转发操作时，路由表是一个非常关键的指引，如果网络中的路由信息出现问题，那么设备的数据转发过程势必受到影响。路由环路（Routing Loop）是一类在路由部署不恰当或网络规划不合理等情况发生后很容易引发的问题。如果网络中的路由信息不正确，则将导致去往某个目的地的数据包在设备之间不停地被来回转发，从而严重影响设备性能，并且大量消耗网络带宽，影响正常的业务流量，这种问题被称为路由环路问题。路由环路对于网络而言是具有严重危害的，任何一个网络规划、设计或交付人员都应该重视并且严格规避该问题。几乎所有的动态路由协议在协议设计时便考虑了路由环路的规避机制，RIP 也不例外。

1. 路由环路的产生

使用 DVRP 的路由器只是简单地将自己的路由表周期性地通告出去，同时将收到的有效路由加载到路由表中，并通过累加的度量值体现到达目的网络的距离。因此使用 DVRP 的路由器的网络中非常容易出现路由环路现象。

在图 3-30 中，R1 和 R2 都运行了 RIP，当网络完成收敛后，R2 通过 RIP 学习到 1.0.0.0/8 路由。现在 R1 的 GE 0/0/1 端口发生了故障，R1 感知到这个拓扑变化并且立即在路由表中删除 1.0.0.0/8 路由。然而，此时这个拓扑变化 R2 并不知晓，R1 准备在下一个更新周期到来时通告这条路由的不可达情况。但是此时完全有可能出现的一种情况是，在 R1 通告这个更新之前，R2 的更新周期到了，它开始在自己的 GE 0/0/0 端口上发送 Response 报文，该报文中包含 R2 的路由表中的所有路由信息，其中就包括 1.0.0.0/8 路由，且该路由的跳数为 2（R2 自己到达该网段需经过 1 台路由器，因此它将路由更新出去时会将跳数加 1）。R1 收到这个 Response 报文后，发现 1.0.0.0/8 竟然通过 R2 可达且跳数为 2，于是它将 1.0.0.0/8 路由加载到路由表中，如图 3-31 所示，这就出现了路由环路。

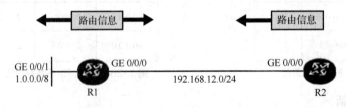

图 3-30 R2 通过 RIP 学习到 1.0.0.0/8 路由

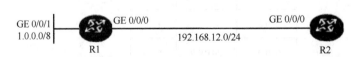

图 3-31 路由环路的产生

路由环路对网络来说危害是非常大的。对于目前的情况，如果 R2 收到一个发往 1.0.0.0/8 的数据包，经过路由表查询后，R2 发现自己有一条路由可以匹配该目的地址，并且下一跳为 R1，因此它将数据包转发给 R1，而 R1 经过路由表查询后，发现到达 1.0.0.0/8 的下一跳是 R2，于是数据包又被转发给 R2，如此反复，发往该网段的数据包就会在 R1、R2 之间不停地被来回转发，直到报文的生存时间（Time To Live，TTL）值递减为 0。可以想象，如果数据流量特别大，这无疑将对路由器的性能造成极大损耗，当然，也可能将链路带宽耗尽。

另外，由于 RIP 每隔 30s 泛洪一次路由表，因此 1.0.0.0/8 会在每个更新周期到来时随着 Response 报文在网络中不停地被泛洪。在 R1 的下一个更新周期到来时，它会把 1.0.0.0/8 路由通过 Response 报文再通告给 R2，R2 收到该报文后，刷新自己的路由表，将该路由的跳数更新为 3，如图 3-32 所示。而当 R2 的更新周期到来时，它将在通告给 R1 的 Response 报文中继续携带 1.0.0.0/8 路由，而且跳数设置为 4。R1 收到该 Response 报文后刷新自己的路由表，将该路由的跳数更新为 4，如此反复。设想一下，如果 RIP 没有任何机制解决该问题，那么 1.0.0.0/8 路由会在网络中被不断地泛洪且其跳数会持续累加到无穷大。

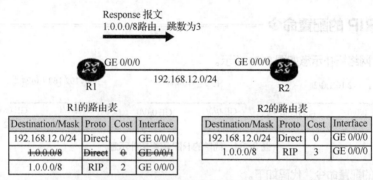

图 3-32　RIP 需防止 1.0.0.0/8 路由在网络中被无休止地传递

因此，从网络设计、协议设计的角度来看都应该充分考虑到路由环路的隐患及可能性，并加以规避。路由环路对于网络的危害性可想而知。为提高性能，防止产生路由环路，RIP 支持水平分割（Split Horizon）与毒性逆转（Poison Reverse），并在毒性逆转时采用触发更新。另外，RIP 还允许引入其他路由协议所得到的路由。

2. RIP 防止路由环路的方法

（1）水平分割

水平分割就是指路由器从某个端口接收的更新信息不允许再从这个端口发回去。也就是说，在路由信息传送过程中，不再把路由信息发送到接收到此路由信息的端口上，从而在一定程度上避免了路由环路的产生。水平分割的防环机制如图 3-33 所示。R2 在收到 R1 和 R3 的路由信息后，在下一次更新时就不会将它们各自的直连网段信息发回去。水平分割可以有效防止路由环路的产生，同时能减少路由更新信息所占用的链路带宽资源。

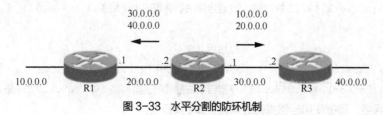

图 3-33　水平分割的防环机制

（2）毒性逆转

毒性逆转可以理解为升级版的水平分割，对于从邻居那里所获得的路由信息，其处理方法不是不发送回去，而是直接将这个网段标记为不可达（16跳）再发送回去，接收方路由器收到路由信息后会立刻抛弃该路由，而不是等待其生存时间到，这样可以加速路由的收敛。毒性逆转会增加路由信息，占用链路带宽，通常不建议使用。

（3）触发更新

若网络中没有变化，则路由器按通常的30s间隔时间发送更新信息。但若有变化，则路由器立即发送其新的路由表。这个过程称为触发更新。

（4）抑制计时

抑制计时即一条路由信息无效之后，一段时间内这条路由都处于抑制状态，即在一定时间内不再接收关于同一目的地址的路由更新。如果路由器从一个网段上得知一条路由失效，然后立即在另一个网段上得知这个路由有效，那么这个有效的信息往往是不正确的，抑制计时就避免了这个问题。此外，当一条链路频繁启停时，抑制计时减少了路由的浮动，增加了网络的稳定性。

3.2.5 RIP 的配置命令

简单的 RIP 网络拓扑示意如图 3-34 所示。

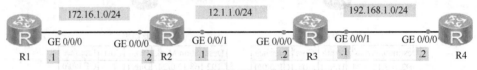

图 3-34　简单的 RIP 网络拓扑示意

（1）RIPv1 的配置命令，代码如下。

```
[R1]rip                          //启用 RIP
[R1-rip-1]version 1              //配置路由器使用 RIPv1，默认是 RIPv1
[R1-rip-1]network 172.16.0.0     //宣告路由器邻居主类网络 172.16.0.0/16 进入 RIP

[R2]rip                          //启用 RIP
[R2-rip-1]version 1              //配置路由器使用 RIPv1，默认是 RIPv1
[R2-rip-1]network 12.0.0.0       //宣告路由器邻居主类网络 12.0.0.0/8 进入 RIP
[R2-rip-1]network 172.16.0.0     //宣告路由器邻居主类网络 172.16.0.0/16 进入 RIP

[R3]rip                          //启用 RIP
[R3-rip-1]version 1              //配置路由器使用 RIPv1，默认是 RIPv1
[R3-rip-1]network 12.0.0.0       //宣告路由器邻居主类网络 12.0.0.0/8 进入 RIP
[R3-rip-1]network 192.168.1.0    //宣告路由器邻居主类网络 192.168.1.0/24 进入 RIP

[R4]rip                          //启用 RIP
[R4-rip-1]version 1              //配置路由器使用 RIPv1，默认是 RIPv1
[R4-rip-1]network 192.168.1.0    //宣告路由器邻居主类网络 192.168.1.0/24 进入 RIP
```

所谓主类网络，是指使用自然掩码的网段。举例如下。

```
12.1.1.0 /24----->12.0.0.0
192.168.1.0 /24----->192.168.1.0
172.16.1.0 /24----->172.16.0.0
```

（2）RIPv2 的配置命令，代码如下。

```
[R1]rip                              //启用RIP
[R1-rip-1]version 2                  //配置路由器使用 RIPv2，默认是 RIPv1
[R1-rip-1]network 172.16.1.0         //宣告路由器邻居网络 172.16.1.0/16 进入 RIP

[R2]rip                              //启用RIP
[R2-rip-1]version 2                  //配置路由器使用 RIPv2，默认是 RIPv1
[R2-rip-1]network 12.1.1.0           //宣告路由器邻居网络 12.1.1.0/8 进入 RIP
[R2-rip-1]network 172.16.1.0         //宣告路由器邻居网络 172.16.1.0/16 进入 RIP

[R3]rip                              //启用RIP
[R3-rip-1]version 2                  //配置路由器使用 RIPv2，默认是 RIPv1
[R3-rip-1]network 12.1.1.0           //宣告路由器邻居网络 12.1.1.0/8 进入 RIP
[R3-rip-1]network 192.168.1.0        //宣告路由器邻居网络 192.168.1.0/24 进入 RIP

[R4]rip                              //启用RIP
[R4-rip-1]version 2                  //配置路由器使用 RIPv2，默认是 RIPv1
[R4-rip-1]network 192.168.1.0        //宣告路由器邻居网络 192.168.1.0/24 进入 RIP
```

3.2.6 OSPF 协议概述

OSPF 协议是一种基于链路状态的内部网关路由协议，运行 OSPF 协议的路由器之间交互的是链路状态信息，而不是直接交互路由。OSPF 将链路状态公告（Link State Advertisement，LSA）传送给在某一区域内的所有路由器，OSPF 路由器将网络中的链路状态信息收集起来，并存储在 LSDB 中。网络中的路由器都有相同的 LSDB，也就是相同的网络拓扑结构，每台 OSPF 路由器都采用 SPF 算法计算到达各个网段的最短路径，并将这些最短路径形成的路由加载到路由表中。

1. OSPF 分层思想

把一个大型网络分割成多个小型网络的能力被称为 OSPF 分层路由，这些被分割出来的小型网络就称为区域。由于区域内部路由器仅与同区域的路由器交换 LSA 信息，这样 LSA 报文数量及链路状态信息库表项都会极大减少，SPF 算法计算速度因此得到提高。多区域的 OSPF 必须存在一个主干区域，主干区域负责收集非主干区域发出的汇总路由信息，并将这些信息返还给各区域。

2. OSPF 区域

每一个网段必须属于一个区域且只能属于一个区域，即每个运行 OSPF 协议的端口必须指定属于某一个特定区域。区域用区域号来标识，区域号是一个从 0 开始的 32 位整数。主干区域不能被非主干区域分割，非主干区域必须和主干区域相连，不建议使用虚连接。

OSPF 主干区域与非主干区域如图 3-35 所示。其中，主干区域是 Area 0 且唯一；非主干区域是非 Area 0 区域。

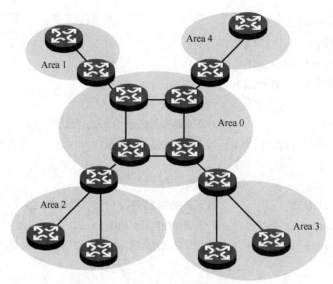

图 3-35　OSPF 主干区域与非主干区域

（1）主干区域

主干区域有且只有一个，其余区域必须与主干区域相连，若没有相连，则无法确定其余区域的路由。全部区域间的路由都必须通过主干区域传递，区域边界路由器（Area Border Router，ABR）必须连接主干区域。主干区域不能被分割，一旦被分割，就必须经过 Vlink 或通用路由封装（Generic Routing Encapsulation，GRE）隧道方式连接。

（2）非主干区域

非主干区域也是通用区域。非主干区域主要承载 Vlink，传输区域内路由、区域间路由和外部路由。

3. OSPF 的网络类型

根据数据链路层协议类型，OSPF 的网络类型分为广播、点到多点、点到点和非广播-多路访问等。

（1）广播类型

当数据链路层协议是以太网或光纤分布式数据接口（Fiber Distributed Data Interface，FDDI）时，OSPF 默认网络类型是广播（Broadcast）类型。在该类型的网络中，通常以组播形式（组播地址为 224.0.0.5 和 224.0.0.6）发送协议报文。

（2）点到多点类型

没有一种数据链路层协议会被默认为点到多点（Point-to-Multi Point，P2MP）类型，必须是由其他的网络类型强制更改的。常用做法是将非广播-多路访问（Non-Broadcast Multi-Access，NBMA）改为 P2MP。在该类型的网络中，默认情况下，以组播形式（组播地址为 224.0.0.5）发送协议报文。可以根据用户需求，以单播形式发送协议报文。

（3）点到点类型

当数据链路层协议是点到点协议（Point-to-Point Protocol，PPP）或高级数据链路控制（High Level Data Link Control，HDLC）协议时，OSPF 默认网络类型是点到点（Point-to-Point，P2P）类型。在该类型的网络中，以组播形式（组播地址为 224.0.0.5）发送协议报文。

（4）非广播-多路访问类型

当数据链路层协议是帧中继、X.25 时，默认情况下，OSPF 认为网络类型是 NBMA 类型。在该类型的网络中，以单播形式发送协议报文。

4. OSPF 路由器类型

在 OSPF 多区域网络中，可以按不同的需求设置以下 4 种路由器，如图 3-36 所示。

① 内部路由器（Internal Router，IR）：所有端口在同一区域的路由器，维护一个 LSDB。

② 主干路由器（Backbone Router，BR）：具有连接主干区域端口的路由器。

③ 区域边界路由器（ABR）：具有连接多区域端口的路由器，一般作为一个区域的出口。ABR 为每一个连接的区域建立 LSDB，负责将所连接区域的路由摘要信息发送到主干区域，而主干区域上的 ABR 负责将这些信息发送到各个区域。

④ 自治系统边界路由器（Autonomous System Border Router，ASBR）：至少拥有一个连接外部自治网络（如非 OSPF 网络）端口的路由器，负责将非 OSPF 网络信息传入 OSPF 网络。

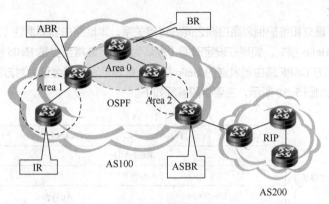

图 3-36　路由器类型

3.2.7　OSPF 报文类型

OSPF 报文主要有 5 种，分别是 Hello 报文、数据库描述（Database Description，DD）报文、链路状态请求（Link State Request，LSR）报文、链路状态更新（Link State Update，LSU）报文、链路状态确认（Link State Acknowledgment，LSAck）报文。这 5 种报文都有相同的报文头，如图 3-37 所示。

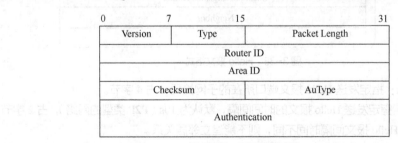

图 3-37　OSPF 报文头

OSPF 报文头主要字段解释如下。

① Version：OSPF 的版本号，占 1 字节。对 OSPFv2 来说，其值为 2。目前最高版本为 OSPFv4，即值为 4（对应二进制数 0100）。

② Type：OSPF 报文的类型，数值从 1～5 分别对应 Hello 报文、DD 报文、LSR 报文、LSU 报文和 LSAck 报文。

③ Packet Length：OSPF 报文的总长度，包括报文头在内，单位为字节。
④ Router ID：路由器 ID 字段，占 4 字节，指定发送报文的源路由器 ID。
⑤ Area ID：区域 ID 字段，占 4 字节，指定发送报文的路由器所对应的 OSPF 区域号。
⑥ Checksum：校验和字段，占 2 字节，是对整个报文（包括 OSPF 报头和各报文具体内容，但不包括下面的 Authentication 字段）的校验和，用于对端路由器校验报文的完整性和正确性。
⑦ AuType：认证类型字段，占 2 字节，指定所采用的认证类型。0 表示不认证；1 表示进行简单认证；2 表示采用 MD5 方式认证。
⑧ Authentication：认证字段，占 8 字节，具体值根据不同认证类型而定。认证类型为不认证时，此字段没有数据；认证类型为简单认证时，此字段为认证密码；认证类型为 MD5 认证时，此字段为 MD5 摘要消息。

1. Hello 报文

Hello 报文用来建立和维护相邻路由器之间的连接关系，本地路由器周期性（默认为 10s）地向邻居路由器端口发送 Hello 报文。如果在设定时间（默认为 40s，通常至少是 Hello 报文发送时间间隔的 4 倍）内没有收到对方 OSPF 路由器发送的 Hello 报文，则本地路由器会认为对方路由器无效。

Hello 报文格式如图 3-38 所示，主要字段解释如下。

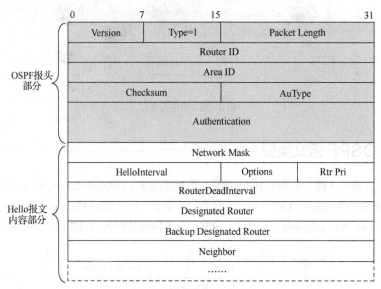

图 3-38 Hello 报文格式

① Network Mask：指定发送 Hello 报文端口所在的子网掩码，占 4 字节。
② HelloInterval：指定发送 Hello 报文的时间间隔，默认为 10s（P2P 类型的网络），占 2 字节。如果相邻两台路由器的 Hello 报文间隔时间不同，则不能建立邻居关系。
③ Options：可选字段，包括 E（允许泛洪 AS-external-LSA）、MC（允许转发 IP 组播报文）、N/P（允许处理 Type-7 LSA），以及 DC（允许处理按需链路），共占 1 字节。
④ Rtr Pri：指定路由器（Designated Router，DR）优先级，默认为 1，占 1 字节。如果设为 0，则表示本路由器不参与 DR/备份指定路由器（Backup Designated Router，BDR）选举。
⑤ RouterDeadInterval：指定路由器失效时间，默认为 40s，占 4 字节。如果在此时间内没有收到邻居路由器发来的 Hello 报文，则认为该邻居路由器已失效。如果相邻两台路由器的失效时间不同，则

不能建立邻居关系。

⑥ Designated Router：指定 DR 的端口 IP 地址，占 4 字节。

⑦ Backup Designated Router：指定 BDR 的端口 IP 地址，占 4 字节。

⑧ Neighbor：指定邻居路由器的路由器标识符（Router Identifier，RID）。其下面的省略号表示可以指定多个邻居路由器的 RID。

2. DD 报文

两台路由器进行数据库同步时，用 DD 报文来描述本地路由器的 LSDB；在两台 OSPF 路由器初始化连接时要交换 DD 报文，进行数据库同步。DD 报文的内容包括 LSDB 中每一条 LSA 的 Header（LSA 的 Header 可以唯一标识一条 LSA）。LSA 的 Header 只占一条 LSA 的整个数据量的一小部分，这样可以减少路由器之间的协议报文流量，对端路由器根据 LSA 的 Header 就可以判断出是否已有这条 LSA。因为数据库的内容可能相当长，所以可能需要多个数据库描述报文对整个数据库进行描述。DD 报文中有 3 个专门用于标识数据库描述报文序列的位数，即 DD 报文格式中的 I、M 和 M/S。接收方对报文的重新排序使其能够真实地复制数据库描述报文。

DD 报文交换过程按询问/应答方式进行，在 DD 报文交换中，一台路由器为主（Master）角色，另一台路由器为从（Slave）角色。Master 路由器向 Slave 路由器发送它的路由表内容，并规定起始序列号，每发送一个 DD 报文，序列号加 1，Slave 路由器则使用 Master 路由器的序列号进行确定应答。但是主从之间的关系会因每个 DD 报文交换的不同而不同。网络中的所有路由器会在不同时刻作用，在这个过程中，其既可能是 Master 路由器，又可能是 Slave 路由器。

DD 报文格式如图 3-39 所示，主要字段解释如下。

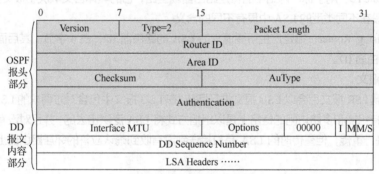

图 3-39　DD 报文格式

① Interface MTU：指定发送 DD 报文的端口在不分段的情况下，可以发出的最大 IP 报文长度，占 2 字节。

② Options：可选字段，包括 E、MC、N/P 及 DC，占 1 字节。

③ I：指定连续发送多个 DD 报文，如果是第一个 DD 报文则置 1，其他均置 0，占 1 比特。

④ M：指定连续发送多个 DD 报文，如果是最后一个 DD 报文则置 0，否则均置 1，占 1 比特。

⑤ M/S：设置 DD 报文交换双方的主从关系，如果本端是 Master 角色，则置 1，否则置 0，占 1 比特。

⑥ DD Sequence Number：指定所发送的 DD 报文序列号，占 4 字节。主从双方利用序列号来确保 DD 报文传输的可靠性和完整性。

⑦ LSA Headers：指定 DD 报文中所包括的 LSA 头部，占 4 字节。其后面的省略号表示可以指定多个 LSA 头部。

因为 DD 报文仅在两台 OSPF 路由器初始化连接时才进行交换，所以它没有发送周期，以后的数据库通过 LSR、LSU 和 LSAck 报文进行同步。

3. LSR 报文

LSR 报文用于请求相邻路由器 LSDB 中的一部分数据。两台路由器互相交换过 DD 报文之后，若想知道对端路由器有哪些 LSA 是本地 LSDB 没有的，以及哪些 LSA 是已经失效的，则需要发送一个 LSR 报文向对方请求所需的 LSA。LSR 报文内容包括所需的 LSA 摘要。

LSR 报文格式如图 3-40 所示，主要字段解释如下。

```
         0              7              15                            31
       ┌──────────────┬──────────────┬──────────────────────────────┐
       │   Version    │   Type=3     │        Packet Length         │
       ├──────────────┴──────────────┴──────────────────────────────┤
OSPF   │                         Router ID                          │
报头   ├────────────────────────────────────────────────────────────┤
部分   │                          Area ID                           │
       ├────────────────────────────┬───────────────────────────────┤
       │         Checksum           │           AuType              │
       ├────────────────────────────┴───────────────────────────────┤
       │                      Authentication                        │
       ├────────────────────────────────────────────────────────────┤
LSR    │                          LS Type                           │
报文   ├────────────────────────────────────────────────────────────┤
内容   │                       Link State ID                        │
部分   ├────────────────────────────────────────────────────────────┤
       │                     Advertising Router                     │
       ├────────────────────────────────────────────────────────────┤
       │                           ……                               │
       └────────────────────────────────────────────────────────────┘
```

图 3-40 LSR 报文格式

① LS Type：指定所请求的 LSA 类型，共 7 类，占 4 字节。

② Link State ID：用于标识网络中的特定路由器或链路，它的具体含义取决于报文类型。在 OSPF 协议中，该字段在不同类型的 LSA 中具有不同的含义。

③ Advertising Router：指定产生所要请求的 LSA 的路由器 ID，占 4 字节。其后面的省略号表示多个 LSA 的路由器 ID。

4. LSU 报文

路由器收到 LSR 报文后会以 LSU 报文进行回应，在 LSU 报文中包含对方请求的 LSA 完整的信息。LSU 报文在支持组播和多路访问的链路上是以组播方式将 LSA 泛洪出去的，并且对没有收到对方确认应答的 LSA 进行重传，但重传时的 LSA 是直接送到没有收到确认应答的邻居路由器上的，而不再是泛洪。

LSU 报文格式如图 3-41 所示，主要字段解释如下。

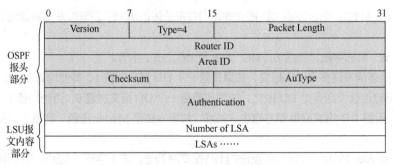

图 3-41 LSU 报文格式

① Number of LSA：指定此报文中共发送的 LSA 数量，占 4 字节。

② LSAs：一条条具体的 LSA 完整信息，占 4 字节。其后面的省略号表示可发送多条 LSA。

5. LSAck 报文

LSAck 报文是路由器在收到对端发来的 LSU 报文后所发出的确认应答报文，内容是需要确认的 LSA 头部（LSA Headers）。LSAck 报文根据不同链路以单播或组播形式发送。LSAck 报文格式如图 3-42 所示。

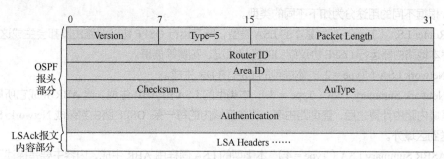

图 3-42　LSAck 报文格式

3.2.8　OSPF 协议的链路状态公告

OSPF 协议是基于链路状态算法的路由协议，所有对路由信息的描述都是封装在 LSA 中发送出去的。当路由器初始化或网络结构发生变化（如增减路由器、链路状态发生变化等）时，路由器会产生 LSA，该数据包中包含路由器所有相连的链路，即所有端口的状态信息。OSPF 路由器收集链路状态信息并使用 SPF 算法计算到各节点的最短路径。

1. LSA 报文

LSA 报文头部格式如图 3-43 所示。所有 LSA 报文头部都有 20 字节，它包含足够的信息来唯一标识一条 LSA。

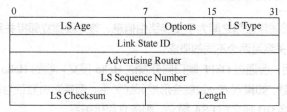

图 3-43　LSA 报文头部格式

LSA 报文头部各字段解释如下。

① LS Age：用来标识链路状态（Link State，LS）产生的时间。生成 LSA 的路由器将 LS 时域初始化为 0，在泛洪过程中，每经过一台路由器，要按 InfTransDelay 的量增加，这个量表示传输 LSA 到下一跳所需要的时间。当该时间达到所设定的 MaxAge 参数时，要撤销该 LSA。

② Options：用来描述支持的路由域，主要包括 DC、EA、N/P、MC、E、T 等选项。DC 指的是始发路由器支持 Demand Circuit（按需拨号等）；EA 指的是始发路由器支持 External。

③ LS Type：链路类型。每种类型的 LSA 都有唯一的通告格式。

④ Link State ID：其内容取决于 LSA 的类型，即不同类型的 LSA 的 Link State ID 是不同的。

⑤ Advertising Router：指始发此 LSA 的路由器的 ID。例如，在 Network LSA 中，这个字段就是 DR 在该网段上端口的 IP 地址。

⑥ LS Sequence Number：用于识别 LSA 报文是不是一个最新包。

⑦ LS Checksum：用于检查 LSA 的完整性，包括除了 LS Age 之外的 LSA 头部的内容。

⑧ Length：LSA 的长度，用字节表示。

2. LSA 的类型

LSA 根据不同的用途分为如下不同的类型。

① Router LSA（Type = 1）：基本的 LSA 类型，所有运行 OSPF 协议的路由器都会生成这种 LSA，主要描述本地路由器运行 OSPF 协议端口的连接状况、花费等信息。

② Network LSA（Type = 2）：本类型的 LSA 由 DR 生成。

③ Network Summary LSA（Type = 3）：本类型的 LSA 由 ABR 生成。当 ABR 完成它所属一个区域中的区域内路由计算之后，查询路由表，将本区域内的每一条 OSPF 路由封装成 Network Summary LSA 发送到区域外。

④ ASBR Summary LSA（Type = 4）：本类型的 LSA 同样由 ABR 生成，内容主要是描述到达本区域内部的 ASBR 的路由。

⑤ AS External LSA（Type = 5）：本类型的 LSA 由 ASBR 生成，主要描述了到自治系统外部路由的信息，LSA 中包含某条路由的目的地址、子网掩码、花费等信息。

⑥ Multicast OSPF LSA（Type = 6）：应用于 OSPF 多播应用程序。

⑦ Not-So-Stubby Area（Type = 7）：本类型的 LSA 由 Not-So-Stubby Area（NSSA）区域中的 ASBR 生成。

⑧ External-Attributes-LSA（Type = 8）：一种特殊的 LSA，还没有实现。

⑨ Opaque LSA（Type = 9～11）：用于多协议标记交换（Multi-Protocol Label Switching，MPLS）流量工程。

3.2.9 邻居和邻接概念的区别

在 OSPF 网络中，邻居路由器之间为了交换路由信息，首先要建立邻接关系。邻居（Neighbor）关系和邻接（Adjacency）关系是两个不同的概念。

OSPF 路由器启动后，会通过 OSPF 端口向外发送 Hello 报文，收到 Hello 报文的 OSPF 路由器会检查报文中所定义的参数，如果双方一致就会形成邻居关系，两端路由器互为邻居。形成邻居关系后，如果两端路由器成功交换 DD 报文和 LSA 报文，则建立邻接关系。

邻居和邻接状态是通过 OSPF 状态机表现的。在 OSPF 邻居建立的过程中，在邻居达到完全邻接关系之前，路由器要经过几个状态。这些状态在 OSPF 标准文档 RFC 2328 中有相关定义，分别是 Down、Attempt、Init、2-Way、ExStart、Exchange、Loading、Full，如图 3-44 所示。Down、2-Way、Full 是稳定状态，Attempt、Init、ExStart、Exchange、Loading 是不稳定状态。不稳定状态是在转换过程中瞬间存在的状态，一般不会超过几分钟。

① Down：双方端口皆为 Down 状态。

② Attempt：此状态仅仅存在于 NBMA 网络手动配置邻居的情况下。在这个状态下，路由器发送单播的 Hello 报文给邻居，且在失效时间超时之前没有收到邻居发过来的 Hello 报文。

③ Init：初始化状态，即单向通信状态，如 A 收到 B 的 Hello 报文（或 B 收到 A 的 Hello 报文）。

④ 2-Way：双方互相通信状态，彼此收到对方的 Hello 报文，并且从 Hello 报文中读取信息，建立邻居关系。

⑤ ExStart：将要开始发送 DD 报文，在发送之前确认谁先发谁后发，自动协商，以路由器 ID 为依据，ID 越大越优先。

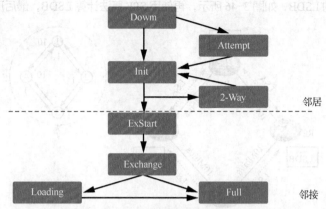

图 3-44　OSPF 邻居到邻接的 8 种状态

⑥ Exchange：确认主从关系后，开始交换 DD 报文（即摘要），会有确认。
⑦ Loading：收到 DD 报文后比较是否相同，并进行 LSR、LSU 的请求和更新。
⑧ Full：若彼此的数据相同（即 LSA 一样），则此状态为邻接关系。

3.2.10　OSPF 工作过程

OSPF 工作过程如图 3-45 所示，简要描述如下。
（1）建立邻居关系，路由器之间会发送 Hello 报文来确认对方是否运行了同一类型的 OSPF 协议。
（2）交互 LSA，并将其存储到 LSDB 中。
（3）路径计算（即 SPF 算法计算）。
（4）生成路由条目。

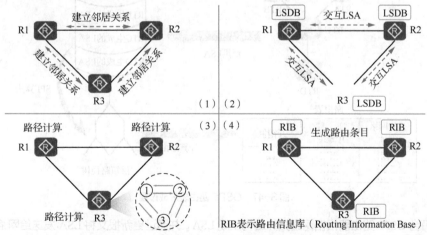

图 3-45　OSPF 工作过程

首先，当路由器启用 OSPF 协议后，路由器之间就会相互发送 Hello 报文，Hello 报文中包含一些路由器和链路的相关信息，发送 Hello 报文的目的是形成邻居表；其次，路由器之间会发送 LSA，LSA

告诉自己的邻居路由器和自己相连的链路的状态；最后，形成网络的拓扑表。其实这个过程是很复杂的，其间经过发送 LSA、记录 LSA、转发 LSA，最后形成 LSDB，即拓扑表。形成拓扑表之后，每台路由器都拥有相同的 LSDB，如图 3-46 所示，再使用 SPF 算法计算 LSDB，最后形成路由表。

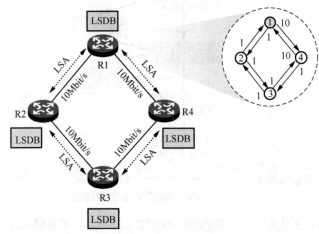

图 3-46　每台路由器都拥有相同的 LSDB

形成路由表后，路由器就可以根据路由表来转发数据包。但是这只是理想情况，如果之后网络拓扑发生了变化，或是网络链路出现了问题，OSPF 协议还是会经过这 3 张表（邻居表、拓扑表、路由表）来重新计算新的路由，只不过不会这么复杂了。在默认情况下，路由器每 10s 就会发送一次 Hello 报文，以检测链路状态，保证链路始终是正常的。

在 OSPF 网络中，OSPF 路由计算过程并不是直接对源地址与目的地址进行关联这么简单，需要考虑到许多因素。OSPF 路由表形成过程如图 3-47 所示，详细步骤如下。

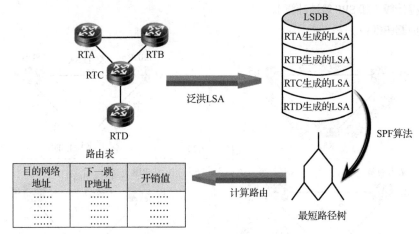

图 3-47　OSPF 路由表形成过程

（1）每台 OSPF 路由器根据网络拓扑结构生成 LSA，并通过更新报文将 LSA 发送给网络中的其他 OSPF 路由器。

（2）每台 OSPF 路由器都会收集其他路由器通告的 LSA，所有的 LSA 组成了 LSDB。LSA 是对路由设备周围网络拓扑结构的描述，LSDB 则是对整个自治系统的网络拓扑结构的描述。

（3）OSPF 路由器将 LSDB 转换成一张带权值的有向图，这张图便是对整个网络拓扑结构的真实反映。各个路由器得到的有向图是完全相同的。

（4）每台路由器根据有向图，使用 SPF 算法计算出一棵以自己为根的最短路径树，这棵树给出了到自治系统中各节点的路由。

3.2.11　OSPF 基本命令

1. OSPF 基本配置命令

（1）ospf 命令

ospf 命令用来创建并运行 OSPF 进程；undo ospf 命令用来关闭 OSPF 进程。默认情况下，系统不运行 OSPF 协议，即不运行 OSPF 进程。

ospf 命令和 undo ospf 命令格式如下。

```
ospf (process-id) router-id (router-id)
undo ospf (process-id)（进程号）
```

其中，process-id 为 OSPF 进程号，采用整数形式，取值为 1~65535，默认值是 1；router-id 为路由器 ID，采用点分十进制形式。

例如，创建并运行 OSPF 进程，进程号为 100，router-id 为 1.1.1.1，代码如下。

```
<Huawei>system-view
Enter system view, return user view with Ctrl+Z.
[Huawei]ospf 100 router-id 1.1.1.1
```

如果要删除进程号为 100 的 OSPF 进程，则代码如下。

```
[Huawei]undo ospf 100
Warning: The OSPF process will be deleted. Continue? [Y/N]:y
[Huawei]
```

（2）area 命令

area 命令用来创建 OSPF 区域，并进入 OSPF 区域视图；undo area 命令用来删除指定 OSPF 区域。默认情况下，系统未创建 OSPF 区域。

area 命令和 undo area 命令格式如下。

```
area (area-id)
undo area (area-id)
```

其中，area-id（即 Area ID）为指定区域的标识，可以是 0，称为主干区域，可以是整数或点分十进制形，采用整数形式时，取值为 0~4294967295。

例如，进入 OSPF 区域视图，代码如下。

```
[Huawei]ospf 100 router-id 1.1.1.1
[Huawei-ospf-100]area 0
[Huawei-ospf-100-area-0.0.0.0]
```

（3）network 命令

network 命令用来指定运行 OSPF 协议的端口和端口所属的区域；undo network 命令用来删除运行 OSPF 协议的端口。默认情况下，此端口不属于任何区域。

network 命令和 undo network 命令格式如下。

```
network (network-address wildcard-mask)
```

```
undo network (network-address wildcard-mask)
```

其中，network-address 表示端口所在的网段地址，采用点分十进制形式；wildcard-mask 为 IP 地址的反码，相当于将 IP 地址的掩码反转（0 变 1，1 变 0），采用点分十进制形式，如 0.0.0.255 表示掩码长度为 24 位。

例如，指定运行 OSPF 协议的端口的主 IP 地址位于网段 192.168.1.0/24，端口所在的 Area ID 为 2，代码如下。

```
[Huawei]ospf 100 router-id 1.1.1.1
[Huawei-ospf-100]area 2
[Huawei-ospf-100-area-0.0.0.2]network 192.168.1.0 0.0.0.255
```

（4）ospf dr-priority 命令

ospf dr-priority 命令用来设置端口在选举 DR 时的优先级；undo ospf dr-priority 命令用来恢复其默认值。默认情况下，其优先级为 1。

ospf dr-priority 命令和 undo ospf dr-priority 命令格式如下。

```
ospf dr-priority (priority)
undo ospf dr-priority
```

其中，priority 表示端口在选举 DR 或 BDR 时的优先级，其值越大，优先级越高，采用整数形式，取值为 0～255。

例如，设置 g0/0/0 端口在选举 DR 时的优先级为 8，代码如下。

```
[Huawei]int g0/0/0
[Huawei-GigabitEthernet0/0/0]ospf dr-priority 8
[Huawei-GigabitEthernet0/0/0]
```

2. 查看 OSPF 基本参数

（1）查看路由器标识

OSPF 路由器标识（Router-ID）是 32 位长度的数值，通常使用点分十进制形式，与 IPv4 地址的格式一样（如 192.168.200.1），用于在 OSPF 区域中唯一地标识一台 OSPF 路由器。一系列连续 OSPF 路由器组成的网络称为 OSPF 区域，这些路由器采用相同的 OSPF 策略。OSPF 要求路由器的 Router-ID 必须全区域唯一，即在同一个区域内不允许出现两台 OSPF 路由器拥有相同的 Router-ID 的情况。

Router-ID 可以使用手动配置的方式进行设定。如果在创建 OSPF 进程时没有手动指定 Router-ID，则系统会自动选择设备上的一个 IP 地址作为 Router-ID。当然，在实际网络部署中，强烈建议大家手动配置 Router-ID，因为这关系到协议的稳定性。一种常见的做法是，将设备的 Router-ID 指定为该设备的 loopback 端口（本地环回端口）的 IP 地址。手动指定 Router-ID（代码中表示为"router-id"）的示例如下。

```
#为设备创建一个 loopback 0 端口，并指定端口的 IP 地址
[Router]interface loopback 0
[Router-LoopBack0]ip address 1.1.1.1 32
[Router-LoopBack0]quit
#创建一个 OSPF 进程，并指定设备的 Router-ID 为 1.1.1.1（loopback 0 端口的 IP 地址）
[Router]ospf 1 router-id 1.1.1.1
```

以上配置中，在系统视图下执行的 ospf 1 router-id 1.1.1.1 命令用于创建一个 OSPF 进程，该进程的 Process-ID 为 1（Process-ID 即进程标识，用于在一台设备上标识一个 OSPF 进程），并且路由器在该

OSPF 进程中所使用的 Router-ID 为 1.1.1.1。无论是采用手动配置还是自动选取的方式，一旦确定了 Router-ID，之后如果再对其进行变更就需要将 OSPF 进程重启，才能使新的 Router-ID 生效。

重启 OSPF 进程的命令如下。

```
<Router>reset ospf process
```

当然，在实际的项目中这条命令需谨慎使用，因为一旦这条命令被执行，OSPF 进程便会重启，该 OSPF 进程的所有邻接关系将会被重置，这会引发路由的动荡。

（2）查看邻居表

在 OSPF 与 LSA 交互之前，两台直连路由器需建立 OSPF 邻居关系。当一个端口激活 OSPF 协议后，该端口将周期性地发送 Hello 报文，同时开始侦听 Hello 报文，从而发现直连链路上的邻居。当 OSPF 协议在端口上发现邻居后，邻居的信息就会被写入路由器的 OSPF 邻居表，随后一个邻接关系的建立过程也就开始了。

在图 3-48 所示的网络中，R1、R2 及 R3 都运行了 OSPF 协议。以 R2 为例，它将在自己的 GE 0/0/0 及 Serial 1/0/0 端口上分别发现 R1、R3，并最终与这两者都建立 OSPF 邻接关系。使用 display ospf peer 命令可查看设备 OSPF 邻居表，使用 display ospf peer brief 命令可查看设备 OSPF 邻居关系简表。以 R2 为例，查看其邻居关系简表的代码和运行结果如下。

```
[R2]display ospf peer brief
        OSPF Process 1 with Router ID 2.2.2.2
              Peer Statistic Information
 ----------------------------------------------------------------------------
 Area Id          Interface                  Neighbor id        State
 0.0.0.0          GigabitEthernet0/0/0       1.1.1.1            Full
 0.0.0.1          Serial 1/0/0               3.3.3.3            Full
```

从上面的结果可看出，R2 已经和 R1、R3 建立了邻居关系。掌握邻居表的查看是使用 OSPF 协议的基本技能之一，也是 OSPF 协议维护及故障定位的重要手段。实际上，每台 OSPF 路由器都与其邻居建立会话，每个会话都使用一个"邻居数据结构"来描述，这些数据结构是与路由器的端口相关联的，它们描述了这个邻居的状态、主从关系、Router-ID、DR 优先级（若有）、端口 IP 地址等信息，OSPF 邻居表则汇总了这些信息，统一将路由器所有邻居的相关数据展示出来。

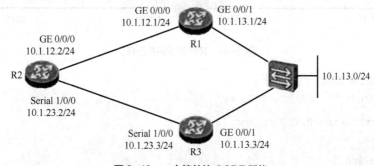

图 3-48　一个简单的 OSPF 网络

（3）查看 LSDB

在 OSPF 网络中，路由器将网络中的 LSA 收集后装载到自己的 LSDB 中，因此 LSDB 可以当作路由器对网络的完整认知。OSPF 协议定义了多种类型的 LSA，这些 LSA 都有各自的用途，但最终的目的都是让路由器知晓网络的拓扑结构及网段信息并计算出最短路径树,从而发现到达全网各个网段的路由。

使用 display ospf lsdb 命令可以查看设备的 LSDB。以 R2 为例,代码如下。

```
[R2]display ospf lsdb
```

实际上,由于 R1、R2 及 R3 的所有端口都属于同一个 OSPF 区域,因此 3 台路由器的 LSDB 都是一致的。

(4)查看路由表

使用 display ospf routing 命令可以查看设备的 OSPF 路由表信息,也就是设备通过 OSPF 协议所发现的路由。以 R2 为例,代码如下。

```
<R2>display ospf routing
```

【任务实施】

实验 2　使用 RIP 配置动态路由

实验目的:

(1)熟悉 RIP 的工作原理。

(2)掌握基本的 RIPv1 配置方法。

(3)理解 RIPv1 与 RIPv2 的区别。

(4)掌握 RIPv2 手动汇总配置方法。

(5)熟悉 RIPv2 认证配置。

RIP 工作过程是在路由器上启用 RIP,将路由器的直连网段宣告进 RIP 后,RIP 会将这些直连网段信息封装成 RIP 报文,发送给邻居路由器,这样邻居路由器间就可以相互学习对方的网段信息,实现网络的互通。配置基本的 RIP 网络主要有两步:启用 RIP;将路由器的直连网段宣告进 RIP。

V3-4　RIPv1 常规配置

1. RIPv1 常规配置

RIP 配置拓扑示意如图 3-49 所示,完成 RIPv1 的常规配置。

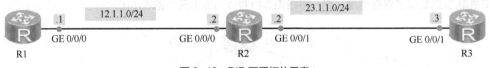

图 3-49　RIP 配置拓扑示意

(1)配置端口 IP 地址,代码如下。

```
[R1]int g0/0/0
[R1-GigabitEthernet0/0/0]ip add 12.1.1.1 24

[R2]int g0/0/0
[R2-GigabitEthernet0/0/0]ip add 12.1.1.2 24
[R2-GigabitEthernet0/0/0]int g0/0/1
[R2-GigabitEthernet0/0/1]ip add 23.1.1.2 24
```

```
[R3]int g0/0/1
[R3-GigabitEthernet0/0/0]ip add 23.1.1.3 24
```

（2）配置 RIP，宣告邻居网段，默认 RIP 为 RIPv1，代码如下。

```
[R1]rip ?
  INTEGER<1-65535>  Process ID
  mib-binding       Mib-Binding a process
  vpn-instance      VPN instance
  <cr>              Please press ENTER to execute command
[R1]rip 100
[R1-RIPV100]network 12.0.0.0


[R2]rip 100
[R2-RIPV100]network 12.0.0.0
[R2-RIPV100]network 23.0.0.0


[R3]rip 100
[R3-RIPV100]network 23.0.0.0
```

（3）查看 RIP 路由信息。查看 R1 的路由表，显示到达 23.0.0.0/8 网络是通过 RIP 获取的，代码如下。

```
[R1]dis ip routing-table
Route Flags: R - relay, D - download to fib
------------------------------------------------------------------------
Routing Tables: Public
        Destinations : 8        Routes : 8
Destination/Mask       Proto    Pre  Cost Flags  NextHop      Interface
12.1.1.0/24            Direct   0    0    D      12.1.1.1     GigabitEthernet 0/0/0
12.1.1.1/32            Direct   0    0    D      127.0.0.1    GigabitEthernet 0/0/0
12.1.1.255/32          Direct   0    0    D      127.0.0.1    GigabitEthernet 0/0/0
23.0.0.0/8             RIP      100  1    D      12.1.1.2     GigabitEthernet 0/0/0
127.0.0.0/8            Direct   0    0    D      127.0.0.1    InLoopBack0
127.0.0.1/32           Direct   0    0    D      127.0.0.1    InLoopBack0
127.255.255.255/32     Direct   0    0    D      127.0.0.1    InLoopBack0
255.255.255.255/32     Direct   0    0    D      127.0.0.1    InLoopBack0
[R1]dis rip 100 route
 Route Flags : R - RIP
             A - Aging, G - Garbage-collect
----------------------------------------------------------------
 Peer 12.1.1.2 on GigabitEthernet0/0/0
     Destination/Mask      Nexthop       Cost    Tag    Flags   Sec
     23.0.0.0/8            12.1.1.2      1       0      RA      15
```

查看 R3 的路由表，显示到达 12.0.0.0/8 网络是通过 RIP 获取的，代码如下。

```
[R3]dis ip routing-table
Route Flags: R - relay, D - download to fib
------------------------------------------------------------------------------
Routing Tables: Public
        Destinations : 8        Routes : 8
Destination/Mask      Proto    Pre  Cost  Flags   NextHop          Interface
12.0.0.0/8            RIP      100  1     D       23.1.1.2         GigabitEthernet 0/0/1
23.1.1.0/24           Direct   0    0     D       23.1.1.3         GigabitEthernet 0/0/1
23.1.1.3/32           Direct   0    0     D       127.0.0.1        GigabitEthernet 0/0/1
23.1.1.255/32         Direct   0    0     D       127.0.0.1        GigabitEthernet 0/0/1
127.0.0.0/8           Direct   0    0     D       127.0.0.1        InLoopBack0
127.0.0.1/32          Direct   0    0     D       127.0.0.1        InLoopBack0
127.255.255.255/32    Direct   0    0     D       127.0.0.1        InLoopBack0
255.255.255.255/32    Direct   0    0     D       127.0.0.1        InLoopBack0
[R3]dis rip 100 route
 Route Flags : R - RIP
             A - Aging, G - Garbage-collect
------------------------------------------------------------------------------
Peer 23.1.1.2 on GigabitEthernet0/0/1
       Destination/Mask         Nexthop       Cost   Tag    Flags   Sec
 12.0.0.0/8                     23.1.1.2      1      0      RA      20
```

（4）测试连通性。由于动态路由 RIP 已启用，因此各路由器之间都有整个网络的路由信息，路由器 R1 与路由器 R3 间是互通的，代码如下。

```
[R1]ping 23.1.1.3
  PING 23.1.1.3: 56  data bytes, press CTRL_C to break
    Reply from 23.1.1.3: bytes=56 Sequence=1 ttl=254 time=40 ms
    Reply from 23.1.1.3: bytes=56 Sequence=2 ttl=254 time=20 ms
    Reply from 23.1.1.3: bytes=56 Sequence=3 ttl=254 time=30 ms
    Reply from 23.1.1.3: bytes=56 Sequence=4 ttl=254 time=30 ms
    Reply from 23.1.1.3: bytes=56 Sequence=5 ttl=254 time=40 ms
 --- 23.1.1.3 ping statistics ---
    5 packet(s) transmitted
    5 packet(s) received
    0.00% packet loss
    round-trip min/avg/max = 20/32/40 ms
[R1]tracert 23.1.1.3
traceroute to  23.1.1.3(23.1.1.3), max hops: 30, packet length: 40, press CTRL_C
 to break
 1 12.1.1.2 20 ms  30 ms  20 ms
 2 23.1.1.3 30 ms  20 ms  20 ms
```

（5）修改参数。RIP 有 3 种计时器，即更新计时器、老化计时器、垃圾回收计时器，可通过 display rip 命令进行查询。更新计时器：运行 RIP 的路由器会以 30s 为周期，向邻居发送 RIP 路由。老化计时器：当路由器在 180s 内都没有收到一条 RIP 路由的更新时，路由器会认为这条 RIP 路由已经失效。垃圾回收计时器：当路由器被设置为失效后，它不会被立即删除，需要等待 120s 才会被彻底删除。

RIP 基于 UDP 发送协议报文，而 UDP 没有确认机制，路由器将路由发送出去以后无法得知邻居路由器是否收到了 RIP 路由。为了保障路由的可靠更新，只有让路由器周期性地发送 RIP 路由（以 30s 为周期），这样从概率上来说能够保证邻居路由器一定可以学习到 RIP 路由。进入 RIP 对计时器做修改时，需要注意保证本地路由老化时间比邻居路由更新时间长。

具体代码如下。

```
//查看RIP进程参数
[R1]dis rip 100
Public VPN-instance
    RIP process : 100    //进程号
    RIP version  : 1     //版本号
    Preference       : 100//优先级
    Checkzero        : Enabled
    Default-cost  : 0
    Summary          : Enabled
    Host-route       : Enabled
    Maximum number of balanced paths : 8
    Update time: 30 sec    //更新时间
    Age time : 180 sec     //老化时间
    Garbage-collect time : 120 sec   //垃圾回收时间
    Graceful restart  : Disabled
    BFD               : Disabled
    Silent-interfaces : None
    Default-route  : Disabled
    Verify-source  : Enabled
    Networks : 12.0.0.0   //宣告网段
//修改RIP计时器
[R1]rip 100
[R1-rip-100]timers ?
  rip  RIP (Routing Information Protocol)
[R1-rip-100]timers rip ?
  INTEGER<1-86400>  Periodic update time (seconds)
[R1-rip-100]timers rip 10 60 40  //修改更新时间为10s,老化时间为60s,垃圾回收时间为40s
[R1-rip-100]
//再次查看RIP进程参数
[R1]dis rip 100
```

```
Public VPN-instance
    RIP process  : 100
    RIP version  : 1
    Preference   : 100
    Checkzero    : Enabled
    Default-cost : 0
    Summary      : Enabled
    Host-route   : Enabled
    Maximum number of balanced paths : 8
    Update time: 10 sec
    Age time : 60 sec
    Garbage-collect time : 40 sec
    Graceful restart : Disabled
    BFD              : Disabled
    Silent-interfaces : None
    Default-route : Disabled
    Verify-source : Enabled
    Networks  : 12
```

（6）设置静默端口。当路由器的端口被设置为 RIP 的静默端口后，这个端口将不会再向外发送 RIP 的组播或广播报文。这样，当路由器对端设备没有启用 RIP 时，可以将本地连接对端设备的端口设置为 RIP 的静默端口。如图 3-50 所示，路由器 R3 不启用 RIP，那么可以将路由器 R2 的 GE 0/0/1 端口设置为静默端口。

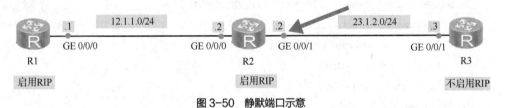

图 3-50　静默端口示意

由于 R3 不启用 RIP，将 R2 的 GE 0/0/1 端口设置为静默端口，代码如下。

```
[R2]rip 100
[R2-rip-100]silent-interface g0/0/1
[R2-rip-100]quit


[R3]undo rip 100
Warning: The RIP process will be deleted. Continue?[Y/N]y
[R3]dis rip
```

在 R2 的 GE 0/0/1 端口进行抓包测试，会发现没有来自 R3 的 RIP 广播报文信息。在 R3 的 GE 0/0/1 端口进行抓包测试，会发现没有来自 R2 的 RIP 广播报文信息。在 R2 的 GE 0/0/0 端口和 R1 的 GE 0/0/0 端口进行测试，会发现有来自 R1 和 R2 的 RIP 广播报文通过。

2. 单播更新配置

所谓单播更新，是指让邻居路由器间通过单播报文交互 RIP 路由信息。当网络内有路由器没有启用 RIP 时，我们就可以让启用了 RIP 的路由器使用单播更新，单独交互 RIP 路由信息，这样就不会影响没有启用 RIP 的路由器了。启用了 RIP 的单播更新功能后，RIP 的组播或广播更新功能并没有关闭，所以需要使用静默端口进行设置，关闭端口的组播和广播更新功能，这样就只有单播更新功能保留下来了。

V3-5 单播更新配置

如图 3-51 所示，假如 R3 不启用 RIP，R1 和 R2 启用 RIP，若不想让 R1 和 R2 上的 RIP 报文被 R3 收到，那么可以通过在 R1 和 R2 上配置单播更新来实现。

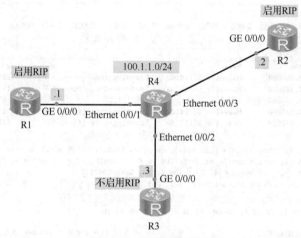

图 3-51 单播更新拓扑示意

（1）配置端口 IP 地址，代码如下。

```
[R1]int g0/0/0
[R1-GigabitEthernet0/0/0]ip add 100.1.1.1 24
[R2]int g0/0/0
[R2-GigabitEthernet0/0/0]ip add 100.1.1.2 24
[R3]int g0/0/0
[R3-GigabitEthernet0/0/0]ip add 100.1.1.3 24
```

（2）配置 RIP。R1 和 R2 启用 RIP，R3 不启用 RIP，代码如下。

```
[R1]rip 1
[R1-rip-1]network 100.0.0.0
[R2]rip 1
[R2-rip-1]network 100.0.0.0
```

（3）配置 RIP 单播更新。

在 R1 上设置单播更新目的地址，设置 GE 0/0/0 端口为静默端口，关闭广播或组播更新功能，代码如下。

```
[R1]rip
[R1-RIPV1]peer 100.1.1.2            //设置单播更新目的地址
[R1-RIPV1]silent-interface g0/0/0   //设置静默端口，关闭广播或组播更新功能
[R1-RIPV1]quit
```

在 R2 上设置单播更新目的地址，设置 GE 0/0/0 端口为静默端口，关闭广播或组播更新功能，代码如下。

```
[R2]rip
[R2-RIPV1]peer 100.1.1.1              //设置单播更新目的地址
[R2-RIPV1]silent-interface g0/0/0 //设置静默端口，关闭广播或组播更新功能
[R2-RIPV1]quit
```

（4）抓包测试。在 R1 的 GE 0/0/0 端口和 R2 的 GE 0/0/0 端口抓包，发现有 RIP 报文信息，如图 3-52 所示。在 R3 的 GE 0/0/0 端口抓包，没有发现 RIP 报文信息，如图 3-53 所示。

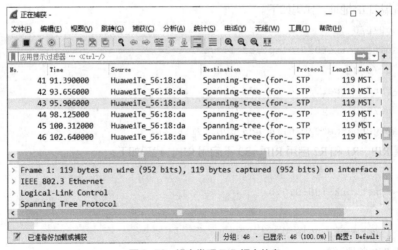

图 3-52　发现有 RIP 报文信息

图 3-53　没有发现 RIP 报文信息

3. RIPv2 配置

相对于 RIPv1 而言，配置 RIPv2 时需要加上 version 2。

使用 RIPv2 完成路由配置的步骤如下。

（1）配置端口 IP 地址，同 RIPv1 配置。

（2）配置 RIPv2，代码如下。

V3-6　RIPv2 配置

```
[R1]rip ?
  INTEGER<1-65535>  Process ID
  mib-binding       Mib-Binding a process
  vpn-instance      VPN instance
  <cr>              Please press ENTER to execute command
[R1]rip 1
[R1-rip-1]version ?
  INTEGER<1-2>  Version of RIP process
[R1-rip-1]version 2
[R1-rip-1]network 12.0.0.0
[R1-rip-1]quit

R2]rip 1
[R2-rip-1]version 2
[R2-rip-1]network 12.0.0.0
[R2-rip-1]network 23.0.0.0
[R2-rip-1]quit

[R3]rip 1
[R3-rip-1]version 2
[R3-rip-1]network 23.0.0.0
[R3-rip-1]
```

（3）查看 RIP 路由信息。在 R1 上查看路由表信息，代码如下。

```
[R1]dis rip 1
Public VPN-instance
   RIP process  : 1          //进程号为1
   RIP version  : 2          //版本号为2
   Preference   : 100        //优先级为100
   Checkzero    : Enabled
   Default-cost : 0
   Summary      : Enabled
   Host-route   : Enabled
   Maximum number of balanced paths : 8
Update time  : 30 sec
Age time : 180 sec
Garbage-collect time : 120 sec
   Graceful restart : Disabled
   BFD              : Disabled
   Silent-interfaces : None
```

```
    Default-route : Disabled
    Verify-source : Enabled
 Networks : 12.0.0.0      //宣告直连网段


[R1]dis ip routing-table
Route Flags: R - relay, D - download to fib
------------------------------------------------------------------------------
Routing Tables: Public
        Destinations : 8        Routes : 8
Destination/Mask    Proto  Pre  Cost  Flags NextHop      Interface
   12.1.1.0/24      Direct  0    0     D   12.1.1.1   GigabitEthernet 0/0/0
   12.1.1.1/32      Direct  0    0     D   127.0.0.1  GigabitEthernet 0/0/0
 12.1.1.255/32      Direct  0    0     D   127.0.0.1  GigabitEthernet 0/0/0
   23.1.1.0/24   RIP   100  1     D   12.1.1.2   GigabitEthernet 0/0/0
  127.0.0.0/8       Direct  0    0     D   127.0.0.1     InLoopBack0
  127.0.0.1/32      Direct  0    0     D   127.0.0.1     InLoopBack0
127.255.255.255/32  Direct  0    0     D   127.0.0.1     InLoopBack0
255.255.255.255/32  Direct  0    0     D   127.0.0.1     InLoopBack0


[R1]dis rip 1 route
Route Flags : R - RIP
            A - Aging, G - Garbage-collect
------------------------------------------------------------------------------
Peer 12.1.1.2 on GigabitEthernet0/0/0
    Destination/Mask       Nexthop     Cost    Tag      Flags   Sec
    23.1.1.0/24           12.1.1.2       1      0        RA     14
```

使用 dis ip routing-table 命令查看路由表时,可以看到前面通过 RIPv1 学到的是 23.0.0.0/8 路由,而通过 RIPv2 学到的是 23.1.1.0/24 路由。其主要原因是 RIPv1 和 RIPv2 都有自动汇总机制,RIPv1 无法关闭自动汇总机制,RIPv2 可以关闭自动汇总机制,并且对于华为设备而言,当启用 RIPv2 时自动汇总机制已经被默认关闭。

所谓自动汇总,是指当子网路由传递到主类网段边界时,路由器会将子网路由汇总为主类网段后再发送出去。

同理,在 R3 上查看路由表信息,可以看到通过 RIPv2 学到的是 12.1.1.0/24 路由,代码如下。

```
[R3]dis ip routing-table
Route Flags: R - relay, D - download to fib
------------------------------------------------------------------------------
Routing Tables: Public
        Destinations : 8        Routes : 8
Destination/Mask    Proto  Pre  Cost  Flags NextHop      Interface
```

```
12.1.1.0/24           RIP     100  1  D  23.1.1.2    GigabitEthernet 0/0/1
23.1.1.0/24           Direct  0    0  D  23.1.1.3    GigabitEthernet 0/0/1
23.1.1.3/32           Direct  0    0  D  127.0.0.1   GigabitEthernet 0/0/1
23.1.1.255/32         Direct  0    0  D  127.0.0.1   GigabitEthernet 0/0/1
127.0.0.0/8           Direct  0    0  D  127.0.0.1   InLoopBack0
127.0.0.1/32          Direct  0    0  D  127.0.0.1   InLoopBack0
127.255.255.255/32    Direct  0    0  D  127.0.0.1   InLoopBack0
255.255.255.255/32    Direct  0    0  D  127.0.0.1   InLoopBack0

[R3]dis rip 1 route
 Route Flags : R - RIP
             A - Aging, G - Garbage-collect
----------------------------------------------------------------

 Peer 23.1.1.2 on GigabitEthernet0/0/1
     Destination/Mask        Nexthop      Cost    Tag    Flags   Sec
    12.1.1.0/24              23.1.1.2     1       0      RA      15
```

（4）路由测试。R1 与 R3 间的互通状态如下。

```
[R1]ping 23.1.1.3
  PING 23.1.1.3: 56  data bytes, press CTRL_C to break
    Reply from 23.1.1.3: bytes=56 Sequence=1 ttl=254 time=30 ms
    Reply from 23.1.1.3: bytes=56 Sequence=2 ttl=254 time=30 ms
    Reply from 23.1.1.3: bytes=56 Sequence=3 ttl=254 time=40 ms
    Reply from 23.1.1.3: bytes=56 Sequence=4 ttl=254 time=20 ms
    Reply from 23.1.1.3: bytes=56 Sequence=5 ttl=254 time=30 ms

 --- 23.1.1.3 ping statistics ---
    5 packet(s) transmitted
    5 packet(s) received
    0.00% packet loss
round-trip min/avg/max = 20/30/40 ms


[R1]tracert 23.1.1.3
 traceroute to  23.1.1.3(23.1.1.3), max hops: 30, packet length: 40, press CTRL_C
to break
 1 12.1.1.2 20 ms  20 ms  10 ms
   23.1.1.3 10 ms  30 ms  30 ms
```

4. RIPv2 手动汇总配置

手动汇总可以减少网络中路由条目的数量，也可以增强网络的稳定性，因为汇总路由中包含的部分明细路由失效是不会影响汇总路由的，只有汇总路由中包含的所有明细路由失效，汇总路由才会消失。RIPv2 可进入端口实现手动汇总，RIPv1 无法实现手动汇总。

如图 3-54 所示，使用手动汇总，将 10.1.0.2/24、10.1.1.2/24、10.1.2.2/24、10.1.3.2/24 这 4 个网段汇总为 10.1.0.0/22 网段。

图 3-54　手动汇总拓扑示意

（1）配置端口 IP 地址，代码如下。

```
[R1]int g0/0/0
[R1-GigabitEthernet0/0/0]ip add 12.1.1.1 24

[R2]int g0/0/0
[R2-GigabitEthernet0/0/0]ip add 12.1.1.2 24
[R2-GigabitEthernet0/0/0]qu
[R2]int loopback 0
[R2-LoopBack0]ip add 10.1.0.2 24
[R2-LoopBack0]int loopback 1
[R2-LoopBack1]ip add 10.1.1.2 24
[R2-LoopBack1]int loopback 2
[R2-LoopBack2]ip add 10.1.2.2 24
[R2-LoopBack2]int loopback 3
[R2-LoopBack3]ip add 10.1.3.2 24
[R2-LoopBack3]qu
[R2]display ip interface brief
*down: administratively down
^down: standby
(l): loopback
(s): spoofing
The number of interface that is UP in Physical is 6
The number of interface that is DOWN in Physical is 2
The number of interface that is UP in Protocol is 6
The number of interface that is DOWN in Protocol is 2
Interface                 IP Address/Mask      Physical     Protocol
GigabitEthernet0/0/0      12.1.1.2/24          up           up
GigabitEthernet0/0/1      unassigned           down         down
GigabitEthernet0/0/2      unassigned           down         down
LoopBack0                 10.1.0.2/24          up           up(s)
LoopBack1                 10.1.1.2/24          up           up(s)
```

LoopBack2	10.1.2.2/24	up	up(s)
LoopBack3	10.1.3.2/24	up	up(s)
NULL0	unassigned	up	up(s)

(2）配置 RIPv2，代码如下。

```
[R1]rip 1
[R1-rip-1]version 2
[R1-rip-1]network 12.0.0.0
[R1-rip-1]quit


[R2]rip 1
[R2-rip-1]version 2
[R2-rip-1]network 12.0.0.0
[R2-rip-1]network 10.0.0.0
[R2-rip-1]quit
```

（3）查看路由信息。查看 R1 学习到的 RIP 路由信息，可以看到 R1 学习到 4 条 RIP 路由信息，代码如下。

```
[R1]dis ip routing-table
Route Flags: R - relay, D - download to fib
------------------------------------------------------------------------
Routing Tables: Public
        Destinations : 11       Routes : 11
Destination/Mask      Proto    Pre  Cost Flags   NextHop         Interface
10.1.0.0/24           RIP      100  1    D       12.1.1.2        GigabitEthernet0/0/0
10.1.1.0/24           RIP      100  1    D       12.1.1.2        GigabitEthernet0/0/0
10.1.2.0/24           RIP      100  1    D       12.1.1.2        GigabitEthernet0/0/0
10.1.3.0/24           RIP      100  1    D       12.1.1.2        GigabitEthernet0/0/0
12.1.1.0/24           Direct   0    0    D       12.1.1.1        GigabitEthernet0/0/0
12.1.1.1/32           Direct   0    0    D       127.0.0.1       GigabitEthernet0/0/0
12.1.1.255/32         Direct   0    0    D       127.0.0.1       GigabitEthernet0/0/0
127.0.0.0/8           Direct   0    0    D       127.0.0.1       InLoopBack0
127.0.0.1/32          Direct   0    0    D       127.0.0.1       InLoopBack0
127.255.255.255/32    Direct   0    0    D       127.0.0.1       InLoopBack0
255.255.255.255/32    Direct   0    0    D       127.0.0.1       InLoopBack0
```

（4）手动汇总。把上面动态获取的 4 条路由信息手动汇总为一条路由信息。在 R2 的端口 GE 0/0/0 上配置 RIPv2 手动汇总，把 10.1.0.0/24、10.1.1.0/24、10.1.2.0/24、10.1.3.0/24 这 4 个网段汇总为 10.1.0.0/22 网段，代码如下。

```
[R2]int g0/0/0
[R2-GigabitEthernet0/0/0]rip sum
[R2-GigabitEthernet0/0/0]rip summary-address 10.1.0.0 ?
  IP_ADDR<X.X.X.X>  IP address mask
```

```
[R2-GigabitEthernet0/0/0]rip summary-address 10.1.0.0 255.255.252.0
```

（5）再次查看路由信息。再次查看 R1 学习到的 RIP 路由信息，发现上面 4 条路由信息已经被汇总为一条路由信息，这样可以简化路由表信息，代码如下。

```
[R1]dis ip routing-table
Route Flags: R - relay, D - download to fib
------------------------------------------------------------------------
Routing Tables: Public
         Destinations : 8        Routes : 8
Destination/Mask        Proto    Pre  Cost  Flags   NextHop         Interface
10.1.0.0/22             RIP      100  1     D       12.1.1.2        GigabitEthernet0/0/0
12.1.1.0/24             Direct   0    0     D       12.1.1.1        GigabitEthernet0/0/0
12.1.1.1/32             Direct   0    0     D       127.0.0.1       GigabitEthernet0/0/0
12.1.1.255/32           Direct   0    0     D       127.0.0.1       GigabitEthernet0/0/0
127.0.0.0/8             Direct   0    0     D       127.0.0.1       InLoopBack0
127.0.0.1/32            Direct   0    0     D       127.0.0.1       InLoopBack0
127.255.255.255/32      Direct   0    0     D       127.0.0.1       InLoopBack0
255.255.255.255/32      Direct   0    0     D       127.0.0.1       InLoopBack0
```

5. RIPv2 认证配置

RIPv2 区别于 RIPv1 的主要特点之一，就是支持明文认证和 MD5 密文认证。通过认证，可以使 RIP 报文广播或组播到指定的邻居。如图 3-55 所示，将 12.1.1.0/24 网段端口设置为明文认证，密码设置为"huawei123"；将 23.1.1.0/24 网段端口设置为密码认证，密码设置为"hw000"。注意：同一段链路上两个端口的认证方式和密码必须保持一致。

V3-7 RIPv2 认证配置

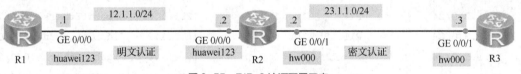

图 3-55 RIPv2 认证配置示意

（1）配置端口 IP 地址，代码如下。

```
[R1]int g0/0/0
[R1-GigabitEthernet0/0/0]ip add 12.1.1.1 24


[R2]int g0/0/0
[R2-GigabitEthernet0/0/0]ip add 12.1.1.2 24
[R2-GigabitEthernet0/0/0]int g0/0/1
[R2-GigabitEthernet0/0/1]ip add 23.1.1.2 24


[R3]int g0/0/1
[R3-GigabitEthernet0/0/0]ip add 23.1.1.3 24
```

（2）配置 RIPv2，代码如下。

```
[R1]rip 1
[R1-rip-1]version 2
[R1-rip-1]network 12.0.0.0

[R2]rip 1
[R2-rip-1]version 2
[R2-rip-1]network 12.0.0.0
[R2-rip-1]network 23.0.0.0

[R3]rip 1
[R3-rip-1]version 2
[R3-rip-1]network 23.0.0.0
```

（3）配置 RIPv2 明文认证。按照图 3-55，将 R1 到 R2 链路段端口设置为明文认证，密码设置为"huawei123"；将 R2 到 R3 链路段端口设置为密文认证，密码设置为"hw000"，代码如下。

```
//将 R1 到 R2 链路段端口设置为明文认证
[R1]int g0/0/0
[R1-GigabitEthernet0/0/0]rip authentication-mode simple cipher huawei123

[R2]int g0/0/0
[R2-GigabitEthernet0/0/0]rip authentication-mode simple cipher huawei123
//将 R2 到 R3 链路段端口设置为密文认证
[R2]int g0/0/1
[R2-GigabitEthernet0/0/1]rip authentication-mode ?
  hmac-sha256
  md5         MD5 authentication
  simple      Simple authentication
[R2-GigabitEthernet0/0/1]rip authentication-mode md5 ?
  nonstandard  Nonstandard MD5 authentication packet format (IETF)
  usual        Huawei MD5 authentication packet format
[R2-GigabitEthernet0/0/1]rip authentication-mode md5 nonstandard ?
  STRING<1-255>/<20-392>  Plain text/Encrypted text
  cipher                  Encryption type (Cryptogram)
  keychain                Configure keychain authentication
  plain                   Encryption type (Plain text)
[R2-GigabitEthernet0/0/1]rip authentication-mode md5 nonstandard cipher hw000 ?
  INTEGER<1-255>  The value of MD5 authentication key ID
[R2-GigabitEthernet0/0/1]rip authentication-mode md5 nonstandard cipher hw000 1

[R3]int g0/0/1
```

[R3-GigabitEthernet0/0/1]rip authentication-mode md5 nonstandard cipher hw000 1

（4）分析抓包内容。如图 3-56 所示，在 R2 的 GE 0/0/0 端口上抓包，发现获取到 RIPv2 报文，报文认证方式为明文认证，密码为"huawei123"。如图 3-57 所示，在 R2 的 GE 0/0/1 端口上抓包，发现获取到 RIPv2 报文，报文认证方式为密文认证，密码已加密。

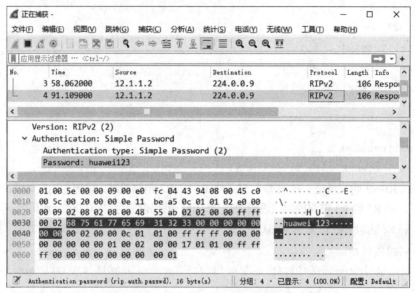

图 3-56　在 R2 的 GE 0/0/0 端口上抓包明文数据

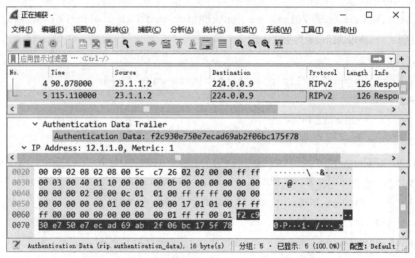

图 3-57　在 R2 的 GE 0/0/1 端口上抓包密文数据

实验 3　使用 OSPF 协议配置动态路由

实验目的：

（1）了解 OSPF 协议的基本特点。

（2）熟悉 OSPF 协议的基本原理与基础配置。

（3）掌握单区域与多区域的 OSPF 配置方法。

（4）熟悉区域认证与链路认证的原理和配置。

（5）了解配置中的相应参数。

划分 OSPF 区域可以缩小路由器的范围，减少流量的使用。区域将路由器划分为不同的组，每个组用 Area ID 标识。每个端口都要指明属于哪一个区域，Area 0 为主干区域，主干区域负责在非主干区域之间发布区域间的路由信息。一个 OSPF 区域中有且只有一个主干区域。设置区域时，子网掩码是反掩码。

V3-8 单区域 OSPF 配置

1. 单区域 OSPF 配置

如图 3-58 所示，两台路由器构成单区域，完成其简单配置。

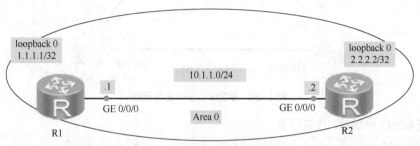

图 3-58 简单 OSPF 拓扑示意

在 R1 和 R2 上分别配置 OSPF，实现动态路由功能的代码如下。

```
[R1]router id 1.1.1.1
[R1]ospf
[R1-ospf-1]area 0
[R1-ospf-1-area-0.0.0.0]network 1.1.1.1 0.0.0.0
[R1-ospf-1-area-0.0.0.0]network 10.1.1.0 0.0.0.255
[R1-ospf-1-area-0.0.0.0]return


[R2]router id 2.2.2.2
[R2]ospf
[R2-ospf-1]area 0
[R2-ospf-1-area-0.0.0.0]network 2.2.2.2 0.0.0.0
[R2-ospf-1-area-0.0.0.0]network 10.1.1.0 0.0.0.255
[R2-ospf-1-area-0.0.0.0]return
[R1]display ospf routing
         OSPF Process 1 with Router ID 1.1.1.1
                 Routing Tables
Routing for Network
Destination        Cost      Type     NextHop         AdvRouter       Area
2.2.2.2/32         1563      Stub     10.1.1.2        2.2.2.2         0.0.0.0
10.1.1.0/24        1562      Stub     10.1.1.1        1.1.1.1         0.0.0.0
1.1.1.1/32         1         Stub     1.1.1.1         1.1.1.1         0.0.0.0
```

实验拓扑示意如图 3-59 所示，完成单区域 OSPF 配置，配置过程如下。

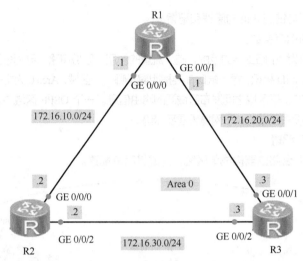

图 3-59 单区域 OSPF 拓扑示意

（1）配置端口 IP 地址，代码如下。

```
[R1]int g0/0/0
[R1-GigabitEthernet0/0/0]ip address 172.16.10.1 24
[R1-GigabitEthernet0/0/0]int g0/0/1
[R1-GigabitEthernet0/0/1]ip address 172.16.20.1 24
[R1-GigabitEthernet0/0/1]

[R2]int g0/0/0
[R2-GigabitEthernet0/0/0]ip address 172.16.10.2 24
[R2-GigabitEthernet0/0/0]int g0/0/2
[R2-GigabitEthernet0/0/2]ip address 172.16.30.2 24
[R2-GigabitEthernet0/0/2]

[R3]int g0/0/1
[R3-GigabitEthernet0/0/1]ip add 172.16.20.3 24
[R3-GigabitEthernet0/0/1]int g0/0/2
[R3-GigabitEthernet0/0/2]ip address 172.16.30.3 24
[R3-GigabitEthernet0/0/2]
```

（2）测试直连链路的连通性，代码如下。

```
[R1-GigabitEthernet0/0/1]ping 172.16.10.2
PING 172.16.10.2: 56  data bytes, press CTRL_C to break
    Reply from 172.16.10.2: bytes=56 Sequence=1 ttl=255 time=60 ms
    Reply from 172.16.10.2: bytes=56 Sequence=2 ttl=255 time=30 ms
    Reply from 172.16.10.2: bytes=56 Sequence=3 ttl=255 time=30 ms
    Reply from 172.16.10.2: bytes=56 Sequence=4 ttl=255 time=20 ms
    Reply from 172.16.10.2: bytes=56 Sequence=5 ttl=255 time=20 ms
```

```
[R1-GigabitEthernet0/0/1]ping 172.16.20.3
[R2-GigabitEthernet0/0/2]ping 172.16.30.3
```

（3）部署单区域 OSPF 网络，代码如下。

```
//配置 R1，将 R1 加入 ospf 1 的 area 0，宣告直连网段
[R1]ospf 1           //启用 OSPF 进程
[R1-ospf-1]area 0    //设置 area 0
[R1-ospf-1-area-0.0.0.0]network 172.16.10.0 0.0.0.255 //宣告直连网段与反掩码
[R1-ospf-1-area-0.0.0.0]network 172.16.20.0 0.0.0.255

//配置 R2，将 R2 加入 ospf 1 的 area 0，宣告直连网段
[R2]ospf 1
[R2-ospf-1]area 0
[R2-ospf-1-area-0.0.0.0]network 172.16.10.0 0.0.0.255
[R2-ospf-1-area-0.0.0.0]network 172.16.30.0 0.0.0.255

//配置 R3，将 R3 加入 ospf 1 的 area 0，宣告直连网段
[R3]ospf 1
[R3-ospf-1]area 0
[R3-ospf-1-area-0.0.0.0]network 172.16.20.0 0.0.0.255
[R3-ospf-1-area-0.0.0.0]network 172.16.30.0 0.0.0.255
```

（4）查看 OSPF 邻居状态，代码如下。

```
[R1]display ospf peer
 OSPF Process 1 with Router ID 172.16.10.1
     Neighbors
Area 0.0.0.0 interface 172.16.10.1(GigabitEthernet0/0/0)'s neighbors
Router ID: 172.16.10.2  Address: 172.16.10.2
  State: Full  Mode:Nbr is Master  Priority: 1
DR: 172.16.10.2  BDR: 172.16.10.1  MTU: 0
  Dead timer due in 30  sec
  Retrans timer interval: 5
  Neighbor is up for 00:05:45
  Authentication Sequence: [ 0 ]
     Neighbors


Area 0.0.0.0 interface 172.16.20.1(GigabitEthernet0/0/1)'s neighbors
Router ID: 172.16.20.3    Address: 172.16.20.3
  State: Full  Mode:Nbr is Master  Priority: 1
DR: 172.16.20.1  BDR: 172.16.20.3  MTU: 0
  Dead timer due in 36  sec
  Retrans timer interval: 5
```

```
    Neighbor is up for 00:05:30
    Authentication Sequence: [ 0 ]
```
通过 Router ID 可以查看邻居的路由器标识；通过 Address 可以查看邻居的 OSPF 端口的 IP 地址；通过 State 可以查看目前与该路由器的 OSPF 邻居状态；通过 Priority 可以查看当前邻居 OSPF 端口的优先级等。

（5）查看路由表，代码如下。

```
[R1]display ip routing-table
Route Flags: R - relay, D - download to fib
------------------------------------------------------------------------
Routing Tables: Public
        Destinations : 11      Routes : 12
Destination/Mask      Proto   Pre Cost Flags NextHop       Interface
127.0.0.0/8           Direct  0   0    D     127.0.0.1     InLoopBack0
127.0.0.1/32          Direct  0   0    D     127.0.0.1     InLoopBack0
127.255.255.255/32    Direct  0   0    D     127.0.0.1     InLoopBack0
172.16.10.0/24        Direct  0   0    D     172.16.10.1   GigabitEthernet0/0/0
172.16.10.1/32        Direct  0   0    D     127.0.0.1     GigabitEthernet0/0/0
172.16.10.255/32      Direct  0   0    D     127.0.0.1     GigabitEthernet0/0/0
172.16.20.0/24        Direct  0   0    D     172.16.20.1   GigabitEthernet0/0/1
172.16.20.1/32        Direct  0   0    D     127.0.0.1     GigabitEthernet0/0/1
172.16.20.255/32      Direct  0   0    D     127.0.0.1     GigabitEthernet0/0/1
172.16.30.0/24        OSPF    10  2    D     172.16.10.2   GigabitEthernet0/0/0
                      OSPF    10  2    D     172.16.20.3   GigabitEthernet0/0/1
255.255.255.255/32    Direct  0   0    D     127.0.0.1     InLoopBack0
```

此时，R1 的路由表中已经拥有了两个去往网络 172.16.30.0/24 的路由条目，代码如下。

```
[R1]display ospf 1 routing
 OSPF Process 1 with Router ID 172.16.10.1
        Routing Tables
 Routing for Network
 Destination        Cost    Type     NextHop       AdvRouter      Area
 172.16.10.0/24     1       Transit  172.16.10.1   172.16.10.1    0.0.0.0
 172.16.20.0/24     1       Transit  172.16.20.1   172.16.10.1    0.0.0.0
 172.16.30.0/24     2       Transit  172.16.10.2   172.16.10.2    0.0.0.0
 172.16.30.0/24     2       Transit  172.16.20.3   172.16.10.2    0.0.0.0
 Total Nets: 4
 Intra Area: 4  Inter Area: 0  ASE: 0  NSSA: 0
```

（6）测试路由。单区域创建完成后，测试路由间的连通性，发现通信正常，代码如下。

```
[R1]tracert 172.16.30.2
traceroute to  172.16.30.2(172.16.30.2), max hops: 30, packet length: 40, press
CTRL_C to break
 1 172.16.20.3 20 ms  20 ms  20 ms
```

```
    2 172.16.10.2 30 ms  20 ms  10 ms
[R1]ping 172.16.30.2
  PING 172.16.30.2: 56  data bytes, press CTRL_C to break
    Reply from 172.16.30.2: bytes=56 Sequence=1 ttl=255 time=20 ms
    Reply from 172.16.30.2: bytes=56 Sequence=2 ttl=255 time=20 ms
    Reply from 172.16.30.2: bytes=56 Sequence=3 ttl=255 time=20 ms
    Reply from 172.16.30.2: bytes=56 Sequence=4 ttl=255 time=20 ms
    Reply from 172.16.30.2: bytes=56 Sequence=5 ttl=255 time=20 ms
  --- 172.16.30.2 ping statistics ---
    5 packet(s) transmitted
    5 packet(s) received
    0.00% packet loss
    round-trip min/avg/max = 20/20/20 ms
```

（7）修改 OSPF 的 RID，代码如下。

```
[R1]ospf 1 router-id 1.1.1.1
Info: The configuration succeeded. You need to restart the OSPF process to valid
ate the new router ID.
<R1>reset ospf process
Warning: The OSPF process will be reset. Continue? [Y/N]:y

[R2]ospf 1 router-id 2.2.2.2
Info: The configuration succeeded. You need to restart the OSPF process to valid
ate the new router ID.
<R2>reset ospf process
Warning: The OSPF process will be reset. Continue? [Y/N]:y

[R3]ospf 1 router-id 3.3.3.3
Info: The configuration succeeded. You need to restart the OSPF process to valid
ate the new router ID.
<R3>reset ospf process
Warning: The OSPF process will be reset. Continue? [Y/N]:y
//查看 R3 路由的邻居，发现是 1.1.1.1 和 2.2.2.2
<R3>display ospf 1 peer
 OSPF Process 1 with Router ID 3.3.3.3
     Neighbors
 Area 0.0.0.0 interface 172.16.30.3(GigabitEthernet0/0/2)'s neighbors
 Router ID: 2.2.2.2         Address: 172.16.30.2
   State: Full  Mode:Nbr is Slave  Priority: 1
   DR: 172.16.30.2  BDR: 172.16.30.3  MTU: 0
   Dead timer due in 30  sec
   Retrans timer interval: 0
```

```
       Neighbor is up for 00:03:04
       Authentication Sequence: [ 0 ]
           Neighbors
 Area 0.0.0.0 interface 172.16.20.3(GigabitEthernet0/0/1)'s neighbors
 Router ID: 1.1.1.1          Address: 172.16.20.1
   State: Full  Mode:Nbr is Slave  Priority: 1
   DR: 172.16.20.1  BDR: 172.16.20.3  MTU: 0
   Dead timer due in 36 sec
   Retrans timer interval: 0
   Neighbor is up for 00:03:11
   Authentication Sequence: [ 0 ]
```

 注意　RID 修改后不会立即生效，需要重启 OSPF 进程，此后新的 RID 才会生效。

（8）修改 OSPF 基准带宽。默认基准带宽是 100Mbit/s，计算各带宽的度量值：10Mbit/s 链路的度量值为 100/10=10；100Mbit/s 链路的度量值为 100/100=1；1000Mbit/s 链路的度量值为 100/1000=0.1（因为度量值不能为小数，所以其结果也是 1）；10000Mbit/s 链路的度量值为 100/10000=0.01（因为度量值不能为小数，所以其结果也是 1）。在默认情况下，100Mbit/s、1000Mbit/s、10000Mbit/s 链路的度量值都是 1，所以需要调整 OSPF 链路上的基准带宽。

查看 R1 路由。当前网络是 1000Mbit/s 链路，度量值为 1，由于 R1 到达 172.16.30.0/24 网络需要经过两段链路，所以下面显示度量值为 2，具体代码如下。

```
[R1]display ip routing-table protocol ospf
Route Flags: R - relay, D - download to fib
------------------------------------------------------------------------------
Public routing table : OSPF
        Destinations : 1      Routes : 2
OSPF routing table status : <Active>
        Destinations : 1      Routes : 2
Destination/Mask  Proto  Pre  Cost Flags NextHop         Interface
172.16.30.0/24    OSPF   10   2    D     172.16.10.2     GigabitEthernet0/0/0
                  OSPF   10   2    D     172.16.20.3     GigabitEthernet0/0/1
OSPF routing table status : <Inactive>
        Destinations : 0      Routes : 0
//修改 OSPF 基准带宽为 10000Mbit/s
[R1]ospf 1
[R1-ospf-1]bandwidth-reference 10000


[R2]ospf 1
[R2-ospf-1]bandwidth-reference 10000
```

```
[R3]ospf 1
[R3-ospf-1]bandwidth-reference 10000


[R1]display ip routing-table protocol ospf
Route Flags: R - relay, D - download to fib
------------------------------------------------------------------------------
Public routing table : OSPF
        Destinations : 1         Routes : 2
OSPF routing table status : <Active>
        Destinations : 1         Routes : 2
Destination/Mask  Proto    Pre  Cost  Flags  NextHop         Interface
172.16.30.0/24    OSPF     10   20    D      172.16.20.3     GigabitEthernet0/0/1
                  OSPF     10   20    D      172.16.10.2     GigabitEthernet0/0/0

OSPF routing table status : <Inactive>
        Destinations : 0         Routes : 0
```

在每台路由设备上修改 OSPF 基准带宽为 10000Mbit/s。因为当前网络是 1000Mbit/s 链路，度量值为 10000/1000=10，R1 到达 172.16.30.0/24 网络需要经过两段链路，所以上面显示度量值为 20。

（9）设置 OSPF 静默端口。RIP 的静默端口不能发送 RIP 报文，但可以接收 RIP 报文。与 RIP 不同，OSPF 协议的静默端口既不能发送又不能接收 Hello 报文，无法与其他路由器建立邻居关系。当网络中某台路由器没有启用 OSPF 协议时，将运行 OSPF 协议的端口指定为 Silent 状态后，该端口的直连路由仍可以发布出去，但端口的 Hello 报文将被阻塞，端口上无法建立邻居关系。若 R3 路由器不启用 OSPF 协议，那么可以把 R1 的 GE 0/0/1 和 R2 的 GE 0/0/2 端口设置为静默端口，不与 R3 建立邻居关系，不向 R3 发送 Hello 报文，代码如下。

```
[R1]ospf 1
[R1-ospf-1]silent-interface g0/0/1

[R2]ospf 1
[R2-ospf-1]silent-interface g0/0/2
```

如图 3-60 所示，通过抓包分析，发现在 172.16.10.0/24 网段的端口上可以获得 OSPF 协议的 Hello 报文。

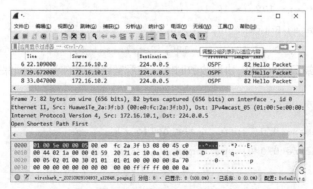

图 3-60　发现 Hello 报文

如图 3-61 所示,此时在 172.16.20.0/24 与 172.16.30.0/24 网段的端口上没有发现 OSPF 协议的 Hello 报文。

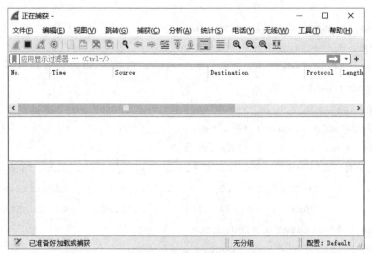

图 3-61 没有发现 Hello 报文

(10) 修改 OSPF 计时。OSPF 协议需要维护邻居表、拓扑表和路由表的更新,这 3 张表的生成时间由于网络类型的不同而不同,其收敛时间也不相同。OSPF 协议主要的网络类型有点到点、点到多点、广播-多路访问、非广播-多路访问。它们的 Hello 时间不全都是 10s,同一物理端口在变更网络类型情况下,其默认 Hello 时间会随之改变,如点到点、广播-多路访问的 Hello 时间默认为 10s,每 10s 发送一次 Hello 报文,Dead 时间默认为 40s,若 40s 都没有收到来自邻居的 Hello 报文,则认为邻居不可达;非广播-多路访问、点到多点的 Hello 时间默认为 30s,Dead 时间默认为 40s。Cisco、华为默认 Hello 时间是 10s,Dead 时间是 Hello 时间的 4 倍,默认为 40s。

在路由端口上查看 OSPF 协议的默认时间,代码如下。

```
[R2]display ospf 1 interface GigabitEthernet 0/0/0
 OSPF Process 1 with Router ID 2.2.2.2
    Interfaces
 Interface: 172.16.10.2 (GigabitEthernet0/0/0)
 Cost: 10      State: BDR      Type: Broadcast    MTU: 1500
 Priority: 1
 Designated Router: 172.16.10.1
 Backup Designated Router: 172.16.10.2
 Timers: Hello 10, Dead 40, Poll 120, Retransmit 5, Transmit Delay 1
```

修改每条链路中端口的 OSPF 协议收敛时间,注意链路两端的端口必须保持一致的收敛时间,代码如下。

```
[R1]int g0/0/0
[R1-GigabitEthernet0/0/0]ospf timer hello 20
[R1-GigabitEthernet0/0/0]ospf timer dead 60

[R2]int g0/0/0
```

```
[R2-GigabitEthernet0/0/0]ospf timer hello 20
[R2-GigabitEthernet0/0/0]ospf timer dead 60
```

//再次查看 OSPF 协议的收敛时间，发现 Hello 和 Dead 时间已修改
```
[R1]dis ospf 1 interface GigabitEthernet 0/0/0
 OSPF Process 1 with Router ID 1.1.1.1
     Interfaces
 Interface: 172.16.10.1 (GigabitEthernet0/0/0)
 Cost: 10      State: BDR      Type: Broadcast    MTU: 1500
 Priority: 1
 Designated Router: 172.16.10.2
 Backup Designated Router: 172.16.10.1
 Timers: Hello 20, Dead 60, Poll 120, Retransmit 5, Transmit Delay 1
```

（11）修改 OSPF 协议的 DR、BDR 角色。在一个 OSPF 网络中，选举一个路由器作为 DR，所有其他路由器只和它交换整个网络的一些路由更新信息，并由它对邻居路由器发送更新报文，这样可以节省网络流量；再指定一个 BDR，当 DR 出现故障时，BDR 发挥作用，确保网络的可靠性。

查看 R1 和 R2 的 OSPF 邻居，发现 DR:172.16.10.2 与 BDR:172.16.10.1，代码如下。

```
[R1]dis ospf 1 peer
 OSPF Process 1 with Router ID 1.1.1.1
       Neighbors
 Area 0.0.0.0 interface 172.16.10.1(GigabitEthernet0/0/0)'s neighbors
 Router ID: 2.2.2.2        Address: 172.16.10.2
   State: Full Mode:Nbr is Master Priority: 1
   DR: 172.16.10.2 BDR: 172.16.10.1 MTU: 0
   Dead timer due in 46 sec
   Retrans timer interval: 5
   Neighbor is up for 00:04:59
   Authentication Sequence: [ 0 ]

[R2]dis ospf 1 peer
 OSPF Process 1 with Router ID 2.2.2.2
       Neighbors
 Area 0.0.0.0 interface 172.16.10.2(GigabitEthernet0/0/0)'s neighbors
 Router ID: 1.1.1.1        Address: 172.16.10.1
   State: Full Mode:Nbr is Slave Priority: 1
   DR: 172.16.10.2 BDR: 172.16.10.1 MTU: 0
   Dead timer due in 57 sec
   Retrans timer interval: 5
   Neighbor is up for 00:06:12
   Authentication Sequence: [ 0 ]
```

如果把上面的 DR 改为 172.16.10.1，那么需要修改 DR 的优先级。由于 DR 优先级默认为 1，因此可降低 R2 的 GE 0/0/0 端口的优先级为 0，代码如下。

```
[R2]int g0/0/0
[R2-GigabitEthernet0/0/0]ospf dr-priority 0
```

注意　将 DR 的优先级提高是不会有效果的，只有把 DR 的优先级降低（0）才能修改角色。当优先级改为 0 的时候，意味着这台设备不能成为 DR，其他想要成为 DR 的自然可以替换为新的 DR。

查看 R1 邻居状态，发现 DR 已更改为 172.16.10.1，邻居 Router ID 为 2.2.2.2，Priority 为 0，代码如下。

```
[R1]dis ospf peer
  OSPF Process 1 with Router ID 1.1.1.1
        Neighbors
 Area 0.0.0.0 interface 172.16.10.1(GigabitEthernet0/0/0)'s neighbors
 Router ID: 2.2.2.2        Address: 172.16.10.2
   State: Full  Mode:Nbr is Master  Priority: 0
 DR: 172.16.10.1  BDR: None   MTU: 0
   Dead timer due in 49 sec
   Retrans timer interval: 5
   Neighbor is up for 00:04:49
   Authentication Sequence: [ 0 ]

[R1]dis ospf 1 interface g0/0/0
  OSPF Process 1 with Router ID 1.1.1.1
        Interfaces
 Interface: 172.16.10.1 (GigabitEthernet0/0/0)
 Cost: 10      State: DR      Type: Broadcast    MTU: 1500
 Priority: 1
 Designated Router: 172.16.10.1
 Backup Designated Router: 0.0.0.0
 Timers: Hello 20, Dead 60, Poll  120, Retransmit 5, Transmit Delay 1
[R2]dis ospf 1 int g0/0/0
  OSPF Process 1 with Router ID 2.2.2.2
        Interfaces
 Interface: 172.16.10.2 (GigabitEthernet0/0/0)
 Cost: 10      State: DROther  Type: Broadcast    MTU: 1500
 Priority: 0
 Designated Router: 172.16.10.1
```

```
Backup Designated Router: 0.0.0.0
Timers: Hello 20, Dead 60, Poll  120, Retransmit 5, Transmit Delay 1
```

2. 多区域 OSPF 配置与认证

OSPF 协议支持报文认证功能，只有通过认证的报文才能接收，否则将不能正常建立邻居关系。OSPF 协议支持区域认证和链路认证。使用区域认证时，一个区域中所有的路由器在该区域下的认证模式和密码必须一致。OSPF 链路认证相比区域认证更加灵活，可专门针对某个邻居设置单独的认证模式和密码。如果同时配置了链路认证，则优先使用链路认证建立 OSPF 邻居。

V3-9 多区域 OSPF 配置与认证

每种认证方式又分为简单验证模式、MD5 验证模式和 Key chain 验证模式。其中，简单验证模式在数据传递过程中，认证密钥和密钥 ID 都是明文传输的，很容易被截获；MD5 验证模式下的密钥经过 MD5 加密传输，相比简单验证模式更为安全；Key chain 验证模式可以同时配置多个密钥，不同密钥可单独设置生效周期等。

多区域 OSPF 认证示意如图 3-62 所示，完成多区域 OSPF 配置及区域认证或链路认证。

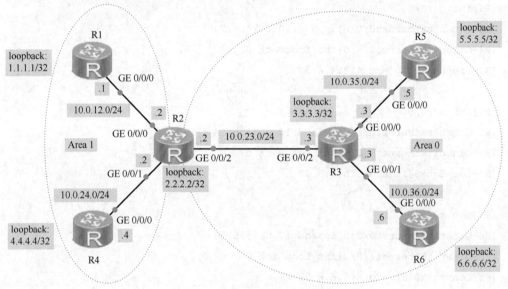

图 3-62 多区域 OSPF 认证示意

（1）配置端口 IP 地址，代码如下。

```
[R1]int LoopBack 0
[R1-LoopBack0]ip add 1.1.1.1 32
[R1-LapBack0]int g0/0/0
[R1-GigabitEthernet0/0/0]ip add 10.0.12.1 24

[R2]int LoopBack 0
[R2-LoopBack0]ip add 2.2.2.2 32
[R2-LoopBack0]int g0/0/0
[R2-GigabitEthernet0/0/0]ip add 10.0.12.2 24
[R2-GigabitEthernet0/0/0]int g0/0/1
```

```
[R2-GigabitEthernet0/0/1]ip add 10.0.24.2 24
[R2-GigabitEthernet0/0/1]int g0/0/2
[R2-GigabitEthernet0/0/2]ip add 10.0.23.2 24

[R3]int LoopBack 0
[R3-LoopBack0]ip add 3.3.3.3 32
[R3-LoopBack0]int g0/0/0
[R3-GigabitEthernet0/0/0]ip add 10.0.35.3 24
[R3-GigabitEthernet0/0/0]int g0/0/1
[R3-GigabitEthernet0/0/1]ip add 10.0.36.3 24
[R3-GigabitEthernet0/0/1]int g0/0/2
[R3-GigabitEthernet0/0/2]ip add 10.0.23.3 24

[R4]int g0/0/0
[R4-GigabitEthernet0/0/0]ip add 10.0.24.4 24
[R4-GigabitEthernet0/0/0]int LoopBack 0
[R4-LoopBack0]ip add 4.4.4.4 32

[R5]int g0/0/0
[R5-GigabitEthernet0/0/0]ip add 10.0.35.5 24
[R5-GigabitEthernet0/0/0]int LoopBack 0
[R5-LoopBack0]ip add 5.5.5.5 32

[R6]int g0/0/0
[R6-GigabitEthernet0/0/0]ip add 10.0.36.6 24
[R6-GigabitEthernet0/0/0]int LoopBack 0
[R6-LoopBack0]ip add 6.6.6.6 32
```

（2）测试直连链路的连通性，代码如下。

```
//在R2与R3上测试整个网络的5条链路的连通性
<R2>ping 10.0.24.4
  PING 10.0.24.4: 56  data bytes, press CTRL_C to break
    Reply from 10.0.24.4: bytes=56 Sequence=1 ttl=255 time=90 ms
    Reply from 10.0.24.4: bytes=56 Sequence=2 ttl=255 time=30 ms
    Reply from 10.0.24.4: bytes=56 Sequence=3 ttl=255 time=40 ms
    Reply from 10.0.24.4: bytes=56 Sequence=4 ttl=255 time=10 ms
    Reply from 10.0.24.4: bytes=56 Sequence=5 ttl=255 time=20 ms
  --- 10.0.24.4 ping statistics ---
    5 packet(s) transmitted
    5 packet(s) received
```

```
    0.00% packet loss
    round-trip min/avg/max = 10/38/90 ms
<R2>ping 10.0.23.3
<R2>ping 10.0.12.1


<R3>ping 10.0.35.5
<R3>ping 10.0.36.3
```

（3）配置区域 OSPF，代码如下。

```
//在 R1 上配置 OSPF，设置 Area 1，宣告直连网段
[R1]ospf
[R1-ospf-1]area 1
[R1-ospf-1-area-0.0.0.1]network 10.0.12.0 0.0.0.255
[R1-ospf-1-area-0.0.0.1]network 1.1.1.1 0.0.0.0
//在 R2 上配置 OSPF，设置 Area 0 和 Area 1，宣告直连网段
[R2]ospf
[R2-ospf-1]area 0
[R2-ospf-1-area-0.0.0.0]network 2.2.2.2 0.0.0.0
[R2-ospf-1-area-0.0.0.0]network 10.0.23.0 0.0.0.255
[R2-ospf-1-area-0.0.0.0]qu
[R2-ospf-1]area 1
[R2-ospf-1-area-0.0.0.1]network 10.0.12.0 0.0.0.255
[R2-ospf-1-area-0.0.0.1]network 10.0.24.0 0.0.0.255
//在 R3 上配置 OSPF，设置 Area 0，宣告直连网段
[R3]ospf
[R3-ospf-1]area 0
[R3-ospf-1-area-0.0.0.0]network 3.3.3.3 0.0.0.0
[R3-ospf-1-area-0.0.0.0]network 10.0.23.0 0.0.0.255
[R3-ospf-1-area-0.0.0.0]network 10.0.35.0 0.0.0.255
[R3-ospf-1-area-0.0.0.0]network 10.0.36.0 0.0.0.255
//在 R4 上配置 OSPF，设置 Area 1，宣告直连网段
[R4]ospf
[R4-ospf-1]area 1
[R4-ospf-1-area-0.0.0.1]network 4.4.4.4 0.0.0.0
[R4-ospf-1-area-0.0.0.1]network 10.0.24.0 0.0.0.255
//在 R5 上配置 OSPF，设置 Area 0，宣告直连网段
[R5]ospf
[R5-ospf-1]area 0
[R5-ospf-1-area-0.0.0.0]network 5.5.5.5 0.0.0.0
[R5-ospf-1-area-0.0.0.0]network 10.0.35.0 0.0.0.255
```

```
//在R6上配置OSPF,设置Area 0,宣告直连网段
[R6]ospf
[R6-ospf-1]area 0
[R6-ospf-1-area-0.0.0.0]network 6.6.6.6 0.0.0.0
[R6-ospf-1-area-0.0.0.0]network 10.0.36.0 0.0.0.0
//每台设备上的环回地址(Loopback)是为了便于测试,所以也要通告到其所在的区域
//配置完成后测试环回端口的连通性,以R1为例
<R1>ping 5.5.5.5
  PING 5.5.5.5: 56  data bytes, press CTRL_C to break
    Reply from 5.5.5.5: bytes=56 Sequence=1 ttl=253 time=30 ms
    Reply from 5.5.5.5: bytes=56 Sequence=2 ttl=253 time=30 ms
    Reply from 5.5.5.5: bytes=56 Sequence=3 ttl=253 time=30 ms
    Reply from 5.5.5.5: bytes=56 Sequence=4 ttl=253 time=40 ms
    Reply from 5.5.5.5: bytes=56 Sequence=5 ttl=253 time=30 ms
  --- 5.5.5.5 ping statistics ---
    5 packet(s) transmitted
    5 packet(s) received
    0.00% packet loss
    round-trip min/avg/max = 30/32/40 ms
```

(4)配置Area 1的OSPF明文认证。在R1上OSPF的Area 1视图下使用authentication-mode命令指定该区域使用认证模式为simple,即简单验证模式,设置密码为"hw123",并设置plain参数。设置plain参数后,可以使得在查看配置文件时,密码均以明文方式显示。如果不设置该参数,则在查看配置文件时,默认会以密文方式显示密码,即无法查看到所配置的密码原文,这可以使非管理员用户在登录设备后无法查看到密码原文,从而提高安全性,具体代码如下。

```
[R1]ospf
[R1-ospf-1]area 1
[R1-ospf-1-area-0.0.0.1]authentication-mode ?
  hmac-md5  Use HMAC-MD5 algorithm
  keychain  Keychain authentication mode
  md5       Use MD5 algorithm
  simple    Simple authentication mode
[R1-ospf-1-area-0.0.0.1]authentication-mode simple ?
  STRING<1-8>/<32>  The password (key)
  cipher            Encryption type (Cryptogram)
  plain             Encryption type (Plain text)
  <cr>              Please press ENTER to execute command
[R1-ospf-1-area-0.0.0.1]authentication-mode simple plain ?
  STRING<1-8>  The password (key)
//设置密码为"hw123"
```

```
[R1-ospf-1-area-0.0.0.1]authentication-mode simple plain hw123
[R1-ospf-1-area-0.0.0.1]dis this
[V200R003C00]
#
 area 0.0.0.1
  authentication-mode simple plain hw123
  network 1.1.1.1 0.0.0.0
  network 10.0.12.0 0.0.0.255
#
return
[R1-ospf-1-area-0.0.0.1]
//如果改为authentication-mode simple hw123，则密码会以密文方式显示
[R1-ospf-1-area-0.0.0.1]authentication-mode simple hw123
[R1-ospf-1-area-0.0.0.1]dis this
[V200R003C00]
#
 area 0.0.0.1
  authentication-mode simple cipher %$%$tqH!"Os\o+;hXo'x-</ZM:SB%$%$
  network 1.1.1.1 0.0.0.0
  network 10.0.12.0 0.0.0.255
#
return
/*使用display ospf peer brief命令查看邻居信息，没有看到R2的信息，说明R1与R2的邻居关系已经中断。因为只是在R1上配置了认证，导致R1与R2间的OSPF认证不匹配*/
[R1]display ospf peer brief
 OSPF Process 1 with Router ID 1.1.1.1
        Peer Statistic Information
 ----------------------------------------------------------------------------
 Area Id         Interface              Neighbor id     State
 ----------------------------------------------------------------------------
//为了查看R2和R4的邻居信息，需要使用同样的方法对R2和R4进行配置，注意密码要一致
[R2]ospf
[R2-ospf-1]area 1
[R2-ospf-1-area-0.0.0.1]authentication-mode simple hw123
[R2-ospf-1-area-0.0.0.1]dis this
[V200R003C00]
#
 area 0.0.0.1
  authentication-mode simple cipher %$%$<xj{%q1)]=<Kxb,_ADR0M~_z%$%$
  network 10.0.12.0 0.0.0.255
```

```
  network 10.0.24.0 0.0.0.255
#
Return

[R4]ospf
[R4-ospf-1]area 1
[R4-ospf-1-area-0.0.0.1]authentication-mode simple hw123
[R4-ospf-1-area-0.0.0.1]dis this
[V200R003C00]
#
 area 0.0.0.1
  authentication-mode simple cipher %$%$:Wx-P$m^@U0|%q%g[=0&MB)6%$%$
  network 4.4.4.4 0.0.0.0
  network 10.0.24.0 0.0.0.255
#
return

[R1]display ospf peer brief
 OSPF Process 1 with Router ID 1.1.1.1
         Neighbors
 Area 0.0.0.1 interface 10.0.12.1(GigabitEthernet0/0/0)'s neighbors
 Router ID: 2.2.2.2          Address: 10.0.12.2
   State: Full   Mode:Nbr is Master  Priority: 1
   DR: 10.0.12.2  BDR: None   MTU: 0
   Dead timer due in 37  sec
   Retrans timer interval: 5
   Neighbor is up for 00:00:02
   Authentication Sequence: [ 0 ]
```

（5）配置 Area 0 区域的 OSPF 密文认证，代码如下。

```
//在 R2 上配置区域认证，验证模式为 MD5，验证字标识符为 1，配置密码为 hw123
[R2]ospf
[R2-ospf-1]area 0
[R2-ospf-1-area-0.0.0.0]authentication-mode md5 1 hw123
[R2-ospf-1-area-0.0.0.0]dis this
[V200R003C00]
#
 area 0.0.0.0
  authentication-mode md5 1 cipher %$%$F_=t"EpoQ0Z&|s-Tfw!'MFKR%$%$
  network 2.2.2.2 0.0.0.0
  network 10.0.23.0 0.0.0.255
```

```
 #
Return
//在R3、R5、R6上配置相同的密文认证
[R3]ospf
[R3-ospf-1]area 0
[R3-ospf-1-area-0.0.0.0]authentication-mode md5 1 hw123

[R5]ospf
[R5-ospf-1]area 0
[R5-ospf-1-area-0.0.0.0]authentication-mode md5 1 hw123

[R6]ospf
[R6-ospf-1]area 0
[R6-ospf-1-area-0.0.0.0]authentication-mode md5 1 hw123
//在R3上查看邻居信息
[R3]dis ospf peer brief
 OSPF Process 1 with Router ID 3.3.3.3
       Peer Statistic Information
 -----------------------------------------------------------------
 Area Id          Interface                  Neighbor id      State
 0.0.0.0          GigabitEthernet0/0/0       10.0.35.5        Full
 0.0.0.0          GigabitEthernet0/0/1       6.6.6.6          Full
 0.0.0.0          GigabitEthernet0/0/2       2.2.2.2          Full
 -----------------------------------------------------------------
```

（6）配置 OSPF 链路认证。在上面的步骤中，使用 OSPF 协议的区域认证方式配置了 OSPF 认证，使用链路认证方式配置可以达到同样的效果。如果采用链路认证的方式，则需要在同一 OSPF 的链路端口下都配置链路认证的命令，设置验证模式和密码等参数；而采用区域认证的方式时，在同一区域中，仅需在 OSPF 进程下的相应区域视图下配置一条命令来设置验证模式和密码即可，大大节省了配置量。所以，在同一区域中如果有多台 OSPF 设备需要配置认证，则建议选用区域认证的方式进行配置。目前已经在 OSPF 区域中配置了简单模式的区域认证，为了进一步提升 R2 与 R4 之间的 OSPF 网络安全性，网络管理员需要在两台设备之间部署 MD5 验证模式的 OSPF 链路认证，具体代码如下。

```
//在R2的 GE 0/0/1端口上配置链路认证，验证模式为MD5，验证字标识符为1，密码为"hw345"
[R2]int g0/0/1
[R2-GigabitEthernet0/0/1]ospf authentication-mode md5 1 hw345
//经过10s，查看邻居信息时发现找不到4.4.4.4。在R4上查看邻居，也看不到任何邻居信息
[R2-GigabitEthernet0/0/1]dis ospf peer br

 OSPF Process 1 with Router ID 2.2.2.2
       Peer Statistic Information
 -----------------------------------------------------------------
```

```
  Area Id          Interface                     Neighbor id        State
  0.0.0.0          GigabitEthernet0/0/2          3.3.3.3            Full
  0.0.0.1          GigabitEthernet0/0/0          1.1.1.1            Full
 -----------------------------------------------------------------------------
[R4]dis ospf peer brief
 OSPF Process 1 with Router ID 10.0.24.4
        Peer Statistic Information
 -----------------------------------------------------------------------------
  Area Id          Interface                     Neighbor id        State
 -----------------------------------------------------------------------------
```
//在 R4 的 GE 0/0/0 端口上配置链路认证，验证模式为 MD5，验证字标识符为 1，口令为"hw345"
```
[R4]int g0/0/0
[R4-GigabitEthernet0/0/0]osp
[R4-GigabitEthernet0/0/0]ospf au
[R4-GigabitEthernet0/0/0]ospf authentication-mode md
[R4-GigabitEthernet0/0/0]ospf authentication-mode md5 1 hw345
[R4-GigabitEthernet0/0/0]dis this
[V200R003C00]
#
interface GigabitEthernet0/0/0
 ip address 10.0.24.4 255.255.255.0
 ospf authentication-mode md5 1 cipher %$%$OgM^"<qZg0ESH>),S!#RMTs!%$%$
#
return
```
//经过 10s，查看邻居信息时发现已找到 2.2.2.2
```
[R4]dis ospf peer br
 OSPF Process 1 with Router ID 10.0.24.4
        Peer Statistic Information
 -----------------------------------------------------------------------------
  Area Id          Interface                     Neighbor id        State
  0.0.0.1          GigabitEthernet0/0/0          2.2.2.2            Full
 -----------------------------------------------------------------------------
```

实验 4　配置 VLAN 间的路由

实验目的：

（1）理解单臂路由的原理和 VLAN 间路由的实现原理。

（2）掌握单臂路由的基本配置方法。

（3）熟悉 VLAN 间的网关设置。

（4）熟悉三层交换机间 VLAN 路由配置。

VLAN 在隔离广播的同时，也禁止了不同 VLAN 间的用户通信，VLAN 间的通信成为新的焦点。VLAN 间路由即交换机的三层路由功能成功地解决了 VLAN 间的通信。

V3-10　单臂路由配置

1. 单臂路由配置

所谓单臂路由，是指利用路由器子端口实现同一交换机上不同 VLAN 间的通信。华为路由器的物理端口可以被划分为多个子端口，每个子端口对应一个 VLAN 的网关。

单臂路由拓扑示意如图 3-63 所示，配置过程如下。

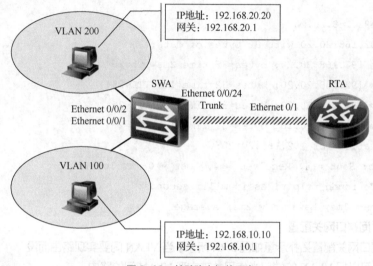

图 3-63　单臂路由拓扑示意

（1）划分 VLAN，代码如下。

```
[SWA]vlan batch 100 200
[SWA]interface Ethernet0/0/1
[SWA-Ethernet0/0/1]port link-type access
[SWA-Ethernet0/0/1]port default vlan 100
[SWA]interface Ethernet0/0/2
[SWA-Ethernet0/0/2]port link-type access
[SWA-Ethernet0/0/2]port default vlan 200
[SWA]int Ethernet0/0/24
[SWA-Ethernet0/0/24]port link-type Trunk
[SWA-Ethernet0/0/24]port Trunk allow-pass vlan all
[SWA-Ethernet0/0/24]quit
```

（2）配置路由器子端口，代码如下。

```
[RTA]interface ethernet 0/1.1          //进入子端口
[RTA-Ethernet0/1.1]control-vid 100 dot1q-termination
[RTA-Ethernet0/1.1]dot1q termination vid 100      //VLAN 100 加入子端口
[RTA-Ethernet0/1.1]arp broadcast enable           //启用 ARP 广播
```

```
[RTA-Ethernet0/1.1]ip address 192.168.10.10 255.255.255.0 //设置子端口IP地址
[RTA]interface ethernet 0/1.2
[RTA-Ethernet0/1.2]control-vid 200 dot1q-termination
[RTA-Ethernet0/1.2]dot1q termination vid 200
[RTA-Ethernet0/1.2]arp broadcast enable
[RTA-Ethernet0/1.2]ip address 192.168.20.20 255.255.255.0
```

（3）测试两台计算机的连通性，代码如下。

```
//测试VLAN 100计算机（IP地址为192.168.10.10）与VLAN 200（IP地址为192.168.20.20）计算机的连通性
C:\>ping 192.168.20.20
Pinging 192.168.20.20 with 32 bytes of data:
Reply from 192.168.20.20: bytes=32 time=2ms TTL=254
Reply from 192.168.20.20: bytes=32 time=1ms TTL=254
Reply from 192.168.20.20: bytes=32 time=1ms TTL=254
Reply from 192.168.20.20: bytes=32 time=1ms TTL=254
Ping statistics for 192.168.20.20:
    Packets: Sent = 4, Received = 4, Lost = 0 (0% loss),
Approximate round trip times in milli-seconds:
    Minimum = 1ms, Maximum = 2ms, Average = 1ms
```

2. VLAN间端口网关配置

VLAN间端口网关配置拓扑示意如图3-64所示。当VLAN间要实现路由而没有路由设备时，可以把VLAN的端口当作网关来实现VLAN间的路由。

V3-11 VLAN间端口网关配置

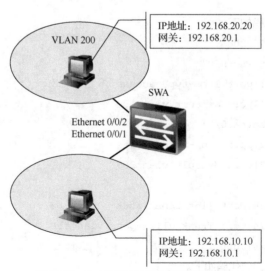

图3-64 VLAN间端口网关配置拓扑示意

（1）划分VLAN，代码如下。

```
[SWA]vlan batch 100 200
[SWA]interface Ethernet0/0/1
```

```
[SWA-Ethernet0/0/1]port link-type access
[SWA-Ethernet0/0/1]port default vlan 100
[SWA]interface Ethernet0/0/2
[SWA-Ethernet0/0/2]port link-type access
[SWA-Ethernet0/0/2]port default vlan 200
```

（2）创建 VLAN 三层端口，代码如下。

```
[SWA]interface Vlanif 100   //进入三层端口
    [SWA-Vlanif100]ip add 192.168.10.10 255.255.255.0 //设置三层端口
IP 地址
    [SWA]interface Vlanif 200
    [SWA-Vlanif200]ip add 192.168.20.20 255.255.255.0
```

V3-12　VLAN 间三层交换机动态路由配置

3. VLAN 间三层交换机动态路由配置

VLAN 间路由拓扑示意如图 3-65 所示，为实现不同 VLAN 间 PC 的通信，要求配置 VLAN 间三层交换机动态路由，配置过程如下。

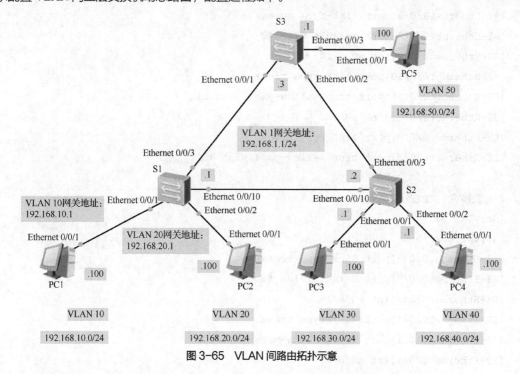

图 3-65　VLAN 间路由拓扑示意

（1）划分 VLAN，代码如下。

```
//在交换机 S1 上划分 VLAN
[S1]vlan batch 10 20
[S1]int e0/0/1
[S1-Ethernet0/0/1]port link-type access
[S1-Ethernet0/0/1]port default vlan 10
[S1-Ethernet0/0/1]int e0/0/2
[S1-Ethernet0/0/2]port link-type access
```

```
[S1-Ethernet0/0/2]port default vlan 20
[S1-Ethernet0/0/2]int e0/0/3
[S1-Ethernet0/0/3]port link-type trunk
[S1-Ethernet0/0/3]port trunk allow-pass vlan all
[S1-Ethernet0/0/3]int e0/0/10
[S1-Ethernet0/0/10]port link-type trunk
[S1-Ethernet0/0/10]port trunk allow-pass vlan all

//在交换机 S2 上划分 VLAN
[S2]vlan batch 30 40
[S2]int e0/0/1
[S2-Ethernet0/0/1]port link-type access
[S2-Ethernet0/0/1]port default vlan 30
[S2-Ethernet0/0/1]int e0/0/2
[S2-Ethernet0/0/2]port link-type access
[S2-Ethernet0/0/2]port default vlan 40
[S2-Ethernet0/0/2]int e0/0/10
[S2-Ethernet0/0/10]port link-type trunk
[S2-Ethernet0/0/10]port trunk allow-pass vlan all
[S2-Ethernet0/0/10]int e0/0/3
[S2-Ethernet0/0/3]port link-type trunk
[S2-Ethernet0/0/3]port trunk allow-pass vlan all

//在交换机 S3 上划分 VLAN
[S3]vlan 50
[S3]int e0/0/3
[S3-Ethernet0/0/3]port link-type access
[S3-Ethernet0/0/3]port default vlan 50
[S3-Ethernet0/0/3]int e0/0/1
[S3-Ethernet0/0/1]port link-type trunk
[S3-Ethernet0/0/1]port trunk allow-pass vlan all
[S3-Ethernet0/0/1]int e0/0/2
[S3-Ethernet0/0/2]port link-type trunk
[S3-Ethernet0/0/2]port trunk allow-pass vlan all
```

(2)配置 VLAN 端口 IP 地址,代码如下。

```
[S1]int Vlanif 1
[S1-Vlanif1]ip add 192.168.1.1 24
[S1-Vlanif1] int Vlanif 10
[S1-Vlanif10]ip add 192.168.10.1 24
[S1-Vlanif10]int vlan 20
```

```
[S1-Vlanif20]ip add 192.168.20.1 24

//交换机 S2 的 VLANIF 地址配置
[S2]int Vlanif 1
[S2-Vlanif1]ip add 192.168.1.2 24
[S2-Vlanif1]int Vlanif 30
[S2-Vlanif30]ip add 192.168.30.1 24
[S2-Vlanif30]int Vlanif 40
[S2-Vlanif40]ip add 192.168.40.1 24

//交换机 S3 的 VLANIF 地址配置
[S3]int Vlanif 1
[S3-Vlanif1]ip add 192.168.1.3 24
[S3-Vlanif1]int Vlanif 50
[S3-Vlanif50]ip add 192.168.50.1 24
```

(3) 动态路由配置，代码如下。

```
[S1]ospf
[S1-ospf-1]area 0
[S1-ospf-1-area-0.0.0.0]network 192.168.10.0 0.0.0.255
[S1-ospf-1-area-0.0.0.0]network 192.168.20.0 0.0.0.255
[S1-ospf-1-area-0.0.0.0]network 192.168.1.0 0.0.0.255

[S1-ospf-1-area-0.0.0.0]dis this
#
 area 0.0.0.0
  network 192.168.10.0 0.0.0.255
  network 192.168.20.0 0.0.0.255
  network 192.168.1.0 0.0.0.255
#
Return

[S2]ospf
[S2-ospf-1]area 0
[S2-ospf-1-area-0.0.0.0]network 192.168.30.0 0.0.0.255
[S2-ospf-1-area-0.0.0.0]network 192.168.40.0 0.0.0.255
[S2-ospf-1-area-0.0.0.0]network 192.168.1.0 0.0.0.255
[S2-ospf-1-area-0.0.0.0]dis this
#
 area 0.0.0.0
  network 192.168.30.0 0.0.0.255
  network 192.168.40.0 0.0.0.255
```

```
  network 192.168.1.0 0.0.0.255
#
Return


[S3]ospf
[S3-ospf-1]area 0
[S3-ospf-1-area-0.0.0.0]network 192.168.50.0 0.0.0.255
[S3-ospf-1-area-0.0.0.0]network 192.168.1.0 0.0.0.255
[S3-ospf-1-area-0.0.0.0]dis this
#
 area 0.0.0.0
  network 192.168.50.0 0.0.0.255
  network 192.168.1.0 0.0.0.255
#
return
```

（4）测试 VLAN 间路由，代码如下。

```
//在 PC5 上分别测试与 PC1、PC2、PC3、PC4 的连通性
PC>ping 192.168.10.100


Ping 192.168.10.100: 32 data bytes, Press Ctrl_C to break
From 192.168.10.100: bytes=32 seq=1 ttl=126 time=79 ms
From 192.168.10.100: bytes=32 seq=2 ttl=126 time=78 ms
From 192.168.10.100: bytes=32 seq=3 ttl=126 time=62 ms
From 192.168.10.100: bytes=32 seq=4 ttl=126 time=78 ms
From 192.168.10.100: bytes=32 seq=5 ttl=126 time=62 ms


--- 192.168.10.100 ping statistics ---
  5 packet(s) transmitted
  5 packet(s) received
  0.00% packet loss
  round-trip min/avg/max = 62/71/79 ms


PC>ping 192.168.20.100
PC>ping 192.168.30.100
PC>ping 192.168.40.100
```

【任务拓展】

根据图 3-66 划分各个 VLAN，厘清每个端口的 IP 地址和各网段的 IP 地址，配置各 PC、服务器和各路由端口的 IP 地址，各直连设备 ping 通之后，配置动态路由（使用 RIP 或 OSPF 协议），实现整个网络的通信。

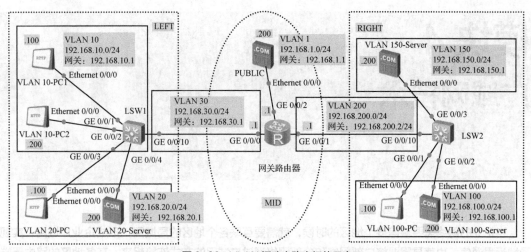

图 3-66 VLAN 间动态路由拓扑示意

模块 4
广域网技术

当企业的网络需要覆盖多个地区的时候,就需要互联各个地区的局域网,形成企业广域网,实现异地远程通信,也就是需要将总部网络和分部网络通过广域网的方式连接起来,在各地区的路由器端口上配置广域网协议实现远距离通信。我们可以在广域网链路上尝试使用 HDLC 协议和 PPP 实现总部与分部广域网的通信,同时 PPP 中的 PAP 认证与 CHAP 认证可以确保企业数据在广域网上通信安全。本模块主要介绍广域网技术,包括 HDLC 协议的配置、PPP 的配置等。

【学习目标】

知识目标
(1)理解广域网的概念及其封装协议。
(2)了解串行链路的数据传输方式。
(3)理解 HDLC 协议、PPP 的概念。
(4)熟悉 HDLC 帧的结构、特点及类型。
(5)熟悉 PPP 帧的结构、特点及认证方式。
(6)理解 PPP 的工作流程。

技能目标
(1)掌握路由器串口的配置方法。
(2)掌握 HDLC 协议和 PPP 的封装配置方法。
(3)掌握 PAP 认证特点和配置方法。
(4)掌握 CHAP 认证特点和配置方法。

培养目标
(1)教育学生制定正确的目标,引导学生制定实现目标的路径。
(2)树立学生正确的人生观、价值观和世界观。
(3)培养学生自主探究问题的精神。
(4)培养学生具备追求卓越的理念,只有持续坚持追求更高的目标,才能不断进步。

任务 4.1 HDLC 协议的配置

【知识准备】

4.1.1 广域网概述

广域网（Wide Area Network，WAN）是一种跨越超大的、地域性的计算机网络集合，通常跨省、市、地区，甚至一个国家。广域网包括很多子网，子网可以是局域网，也可以是小型的广域网。因为串行通信有着传输距离远、成本低的特点，所以远距离、超远距离的通信中较常使用串行通信。串行链路普遍用于广域网中。

典型的串口封装协议有以下两种：高级数据链路控制（HDLC）协议、点到点协议（PPP）。

4.1.2 串行链路的数据传输方式

串行链路中定义了两种数据传输方式，即异步和同步，如图 4-1 所示。

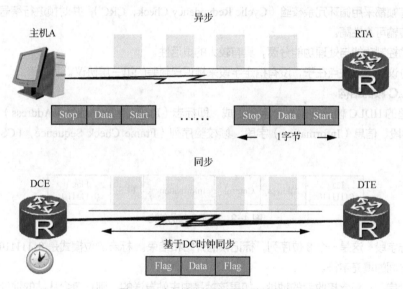

图 4-1 串行链路的两种数据传输方式

异步传输是以字节为单位来传输数据的，并且需要采用额外的起始位和停止位来标记每个字节的开始和结束。起始位为二进制值 0，停止位为二进制值 1。在这种传输方式下，起始位和停止位占据发送数据的一大部分，每个字节的发送都需要额外的开销，可以面向点到点的传输，也可以面向点到多点的传输。

同步传输是以帧为单位来传输数据的，在通信时需要使用时钟来同步本端和对端的设备通信。数据通信设备（Data Circuit-terminating Equipment，DCE）提供了一个用于同步 DCE 和数据终端设备（Data Terminal Equipment，DTE）之间数据传输的时钟信号。其中，DTE 通常使用 DCE 产生的时钟信号。

4.1.3 HDLC 协议

1. HDLC 协议简介

HDLC 协议是一种面向位的数据链路层协议，用以实现远程用户间的资源共享和信息交互。HDLC 协议是数据链路层协议的一项国际标准，用以保证传输到下一层的数据在传输过程中能够准确地被接收，即差错释放中没有任何损失，并且序列正确。HDLC 链路的通信如图 4-2 所示。

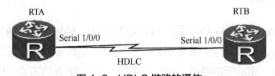

图 4-2 HDLC 链路的通信

HDLC 协议在同步串行数据链路上对数据进行封装，常用于点到点连接的场景中。HDLC 协议主要有以下几个优点。

（1）协议不依赖于任何一种字符编码集。

（2）数据报文可透明传输，用于透明传输的"0 比特插入法"易于硬件实现。

（3）全双工通信，不必等待确认，可连续发送数据报文，有较高的数据链路传输效率。

（4）所有帧都采用循环冗余校验（Cyclic Redundancy Check，CRC），并对帧进行编号，可防止漏收或重收，传输可靠性高。

（5）传输控制功能与处理功能分离，具有较大的灵活性。

HDLC 协议的主要缺点在于，没有指定字段来标识已封装的第三层协议。

2. HDLC 帧的结构

一个完整的 HDLC 帧最多由 6 个字段组成，即标志（Flag）字段、地址（Address）字段、控制（Control）字段、信息（Information）字段、帧校验序列（Frame Check Sequence，FCS）字段，如图 4-3 所示。

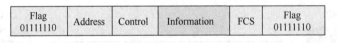

图 4-3 HDLC 帧的结构

（1）标志字段：这是一个 8 位序列，标记帧的开始和结束。标志的位模式是 01111110。它也可以作为帧与帧之间的填充字符。

（2）地址字段：包含接收者的地址。如果该帧是由主站发送的，则它包含从站的地址；如果该帧是由从站发送的，则它包含主站的地址。地址字段的长度可以从 1 字节到几字节。

（3）控制字段：用于构成各种命令及响应，以便对链路进行监控。控制字段的长度为 1 字节或 2 字节。

（4）信息字段：承载来自网络层的数据。它的长度由 FCS 字段或通信节点的缓存容量来决定。使用较多的上限是 1000～2000 比特；下限是 0（S 帧）。

（5）帧校验序列字段：这是一个 2 字节或 4 字节的帧检查序列，用于对两个标志字段之间的内容进行错误检测。使用的是标准代码 CRC。

3. HDLC 帧的类型

根据不同的控制字段，HDLC 帧可分为不同类型，即信息帧（I 帧）、监控帧（S 帧）和无编号帧

（U 帧），如图 4-4 所示。

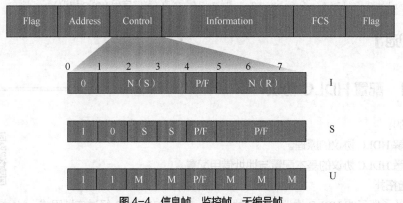

图 4-4　信息帧、监控帧、无编号帧

（1）信息帧：用于承载来自网络层的用户数据，用于传送有效信息或数据，还包括附带在用户数据上的流和错误控制信息。信息帧控制字段的第一位为 0。

（2）监控帧：用于差错控制和流量控制，不包含信息字段。当不需要负载时，该帧用于流量和错误控制。其中，监控帧控制字段的前两位为 10。

（3）无编号帧：用于提供对链路的建立、拆除和多种控制功能，如连接管理（链路的建立与拆除）。如果需要，则还可以包含一个信息字段。其中，无编号帧控制字段的前两位为 11。

4. HDLC 端口地址借用

若端口没有 IP 地址，则无法产生路由，也就无法转发数据报文。IP 地址借用允许一个没有 IP 地址的端口从其他端口上借用 IP 地址，这样可以避免一个端口独占 IP 地址，从而造成 IP 地址的浪费。一般是借用 loopback 端口的 IP 地址。由于这类端口总是处于活跃状态，因此能提供稳定、可用的 IP 地址。

5. HDLC 协议的配置命令

HDLC 协议简单配置拓扑示意如图 4-5 所示。

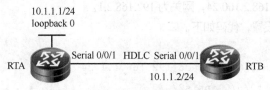

图 4-5　HDLC 协议简单配置拓扑示意

HDLC 协议的具体配置命令如下。

```
[RTA]interface Loop0
[RTA-LoopBack0]ip address 10.1.1.1 24
[RTA]interface Serial 0/0/1
[RTA-Serial0/0/1]link-protocol hdlc      //端口启用 HDLC 协议
[RTA-Serial0/0/1]ip address unnumbered interface LoopBack 0
//IP 地址借用：从其他端口借用 IP 地址，一般建议借用 loopback 端口的 IP 地址
[RTA]ip route-static 10.1.1.2 255.255.255.255 Serial 0/0/1

[RTB]interface Serial 0/0/1
```

```
[RTB-Serial0/0/1]link-protocol hdlc
[RTB-Serial0/0/1]ip address 10.1.1.2 24        //端口启用 HDLC 协议
```

【任务实施】

实验 1　配置 HDLC 协议

实验目的：
（1）了解 HDLC 协议的原理。
（2）熟悉 HDLC 协议的基本配置与地址借用配置。

V4-1　配置 HDLC 协议

1. 实验拓扑

根据图 4-6 所示的 HDLC 协议配置拓扑示意，实现 PC1、PC2 经过广域网或运营商网络的通信，完成广域网串行链路 HDLC 封装协议与 PPP 封装协议配置。

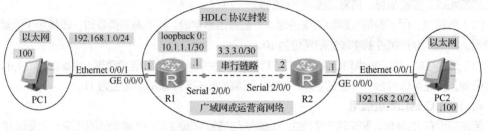

图 4-6　HDLC 协议配置拓扑示意

2. 实验过程

（1）PC 编址，具体如下。

PC1：IP 地址为 192.168.1.100/24，网关为 192.168.1.1。

PC2：IP 地址为 192.168.2.100/24，网关为 192.168.2.1。

（2）路由器 IP 地址设置，代码如下。

```
[R1]int g0/0/0
[R1-GigabitEthernet0/0/0]ip add 192.168.1.1 24
[R1-GigabitEthernet0/0/0]int s2/0/0
[R1-Serial2/0/0]ip add 3.3.3.1 30

[R2]int g0/0/0
[R2-GigabitEthernet0/0/0]ip add 192.168.2.1 24
[R2-GigabitEthernet0/0/0]int s2/0/0
[R2-Serial2/0/0]ip add 3.3.3.2 30
```

（3）串口封装协议设置。华为设备要求串口封装协议一致，默认封装协议为 PPP，可以使用 link-protocol 命令更改封装协议，如配置 HDLC 协议封装。检测串行链路，由于两边封装协议不一样，一边是 HDLC 协议，另一边是 PPP，还需要将 R2 的 Serial 2/0/0 串口改为 HDLC 协议封装，具体代码如下。

```
[R1-Serial2/0/0]dis this
[V200R003C00]
```

```
#
interface Serial2/0/0
link-protocol ppp
 ip address 3.3.3.1 255.255.255.252
#
return
#使用 link-protocol 命令更改封装协议，如配置 HDLC 协议封装
[R1]int s2/0/0
[R1-Serial2/0/0]lin
[R1-Serial2/0/0]link-protocol ?
  fr    Select FR as line protocol
  hdlc  Enable HDLC protocol
  lapb  LAPB(X.25 level 2 protocol)
  ppp   Point-to-Point protocol
  sdlc  SDLC(Synchronous Data Line Control) protocol
  x25   X.25 protocol
[R1-Serial2/0/0]link-protocol hdlc
Warning: The encapsulation protocol of the link will be changed. Continue? [Y/N]:y
Nov  9 2021 11:12:29-08:00 R1 %%01IFNET/4/CHANGE_ENCAP(l)[2]:The user performed
the configuration that will change the encapsulation protocol of the link and th
en selected Y.
[R1]dis interface Serial2/0/0
Serial2/0/0 current state : UP
Line protocol current state : UP
Last line protocol up time : 2021-11-09 11:16:07 UTC-08:00
Description:HUAWEI, AR Series, Serial2/0/0 Interface
Route Port, The Maximum Transmit Unit is 1500, Hold timer is 10(sec)
Internet Address is 3.3.3.1/30
Link layer protocol is nonstandard HDLC
……
#R2 的 Serial 2/0/0 串口改为 HDLC 协议封装
[R2]int s2/0/0
[R2-Serial2/0/0]link-protocol hdlc
Warning: The encapsulation protocol of the link will be changed. Continue? [Y/N]:y
Nov  9 2021 11:15:56-08:00 R2 %%01IFNET/4/CHANGE_ENCAP(l)[0]:The user performed
the configuration that will change the encapsulation protocol of the link and then
selected Y.
```

（4）配置静态路由，实现 PC1 与 PC2 的互通，代码如下。

```
[R1]ip route-static 192.168.2.0 24 Serial 2/0/0
[R2]ip route-static 192.168.1.0 24 Serial 2/0/0
```

此时，PC1 经过广域网能到达 PC2。在 PC1 上 ping PC2 的结果如图 4-7 所示。

图 4-7　在 PC1 上 ping PC2 的结果

（5）地址借用配置。在 R1 上配置 loopback 0 地址为 10.1.1.1/30。在 R1 上删除 Serial 2/0/0 上的 IP 地址 3.3.3.1，使它借用 R1 上的 loopback 0 地址 10.1.1.1，具体代码如下。

```
[R1]int LoopBack 0
[R1-LoopBack0]ip add 10.1.1.1 30
//删除 IP 地址
[R1]int s2/0/0
[R1-Serial2/0/0]undo ip address 3.3.3.1 255.255.255.252
[R1]dis ip int brief
Interface                 IP Address/Mask       Physical      Protocol
GigabitEthernet0/0/0      192.168.1.1/24        up            up
GigabitEthernet0/0/1      unassigned            down          down
GigabitEthernet0/0/2      unassigned            down          down
LoopBack0                 10.1.1.1/30           up            up(s)
NULL0                     unassigned            up            up(s)
Serial2/0/0               unassigned            up            up
Serial2/0/1               unassigned            down          down

//地址借用
[R1]int s2/0/0
[R1-Serial2/0/0]ip add unnumbered interface LoopBack 0

//地址查看
[R1]dis ip int brief
Interface                 IP Address/Mask       Physical      Protocol
GigabitEthernet0/0/0      192.168.1.1/24        up            up
GigabitEthernet0/0/1      unassigned            down          down
```

GigabitEthernet0/0/2	unassigned	down	down
LoopBack0	10.1.1.1/30	up	up(s)
NULL0	unassigned	up	up(s)
Serial2/0/0	10.1.1.1/30	up	up
Serial2/0/1	unassigned	down	down

（6）测试。测试串行链路，发现是不通的。但 PC1 可以 ping 通 PC2。在运营商的主干网或者为 Internet 提供路由服务的链路上，在 IPv4 地址极其紧缺的情况下，地址借用可以节约 IP 地址。使用客户端上网时，没有必要去访问运营商的路由服务器，需要访问的是网络中的资源服务器，具体代码如下。

```
//测试串行链路
[R1]ping 3.3.3.2
  PING 3.3.3.2: 56  data bytes, press CTRL_C to break
    Request time out
    Request time out
    Request time out
    Request time out
    Request time out

//在 PC1 上 ping PC2
PC>ping 192.168.2.100
Ping 192.168.2.100: 32 data bytes, Press Ctrl_C to break
From 192.168.2.100: bytes=32 seq=1 ttl=126 time=15 ms
From 192.168.2.100: bytes=32 seq=2 ttl=126 time=16 ms
From 192.168.2.100: bytes=32 seq=3 ttl=126 time=15 ms
From 192.168.2.100: bytes=32 seq=4 ttl=126 time=16 ms
From 192.168.2.100: bytes=32 seq=5 ttl=126 time<1 ms
```

【任务拓展】

HDLC 协议封装如图 4-8 所示，模拟广域网路由器互联，路由器 RA 与路由器 RB 分别作为 DCE 端与 DTE 端，广域网链路采用 HDLC 协议封装，根据拓扑完成广域网链路的 HDLC 协议的封装配置及路由器串口配置。

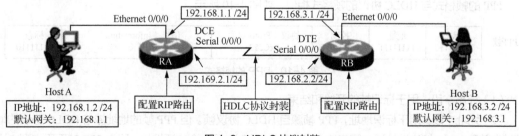

图 4-8 HDLC 协议封装

任务 4.2 PPP 的配置

【知识准备】

4.2.1 PPP 概述

PPP 是一种点到点的数据链路层协议，主要用于在全双工的同步/异步链路上进行点到点的数据传输，并按照顺序传输数据包。PPP 链路的通信如图 4-9 所示。

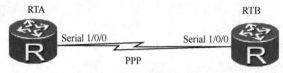

图 4-9 PPP 链路的通信

PPP 是目前使用广泛的数据链路层协议，不管是低速的"拨号猫"连接还是高速的光纤链路，都适用 PPP。Internet 用户通常要连接到某个 ISP 才能接入 Internet。PPP 就是用户计算机和 ISP 进行通信时所使用的数据链路层协议。ISP 使用 PPP 为计算机分配一些网络参数（如 IP 地址、域名等）。

PPP 更像是一种应用，类似于一款拨号上网的应用软件，拨号成功后，本地主机就可以正常上网，可以使用 TCP/IP，而完全感觉不到 PPP 的存在。PPP 是一个协议集，主要包含链路控制协议（Link Control Protocol，LCP）、网络控制协议（Network Control Protocol，NCP）和一些 PPP 的扩展协议，如多链路协议（Multilink Protocol，MP）等。PPP 的主要优点如下。

（1）PPP 既支持同步传输又支持异步传输。而 X.25、帧中继等数据链路层协议仅支持同步传输，串行线路网际协议（Serial Line Internet Protocol，SLIP）仅支持异步传输。

（2）PPP 具有很好的扩展性，在以太网中时，可以扩展为基于以太网的点到点协议（Point-to-Point Protocol over Ethernet，PPPoE）。

（3）PPP 提供了 LCP 用于数据链路层参数的协商，提供了 NCP 用于网络层参数的协商。

（4）PPP 提供了挑战握手身份认证协议（Challenge Handshake Authentication Protocol，CHAP）、密码认证协议（Password Authentication Protocol，PAP），更好地保证了网络的安全性。

（5）PPP 无重传机制，网络开销小，速度快。

4.2.2 PPP 的帧格式

PPP 的帧格式与 HDLC 协议的帧格式相似，如图 4-10 所示。

PPP帧	标志 01111110	地址 11111111	控制 00000011	协议（Protocol） 0xC223	信息（Information） 变长	标志 01111110

图 4-10 PPP 的帧格式

（1）标志字段：用于标识帧的开始和结束。

（2）地址字段：用于标识地址，PPP 帧源自 HDLC 协议帧。但 PPP 帧的地址域与 HDLC 协议的地址域有差异，PPP 帧的地址域为固定的 11111111（0xFF），是一个广播地址，因为 PPP 帧采用点到点协议，不需要地址位。

（3）控制字段：在 HDLC 协议的帧中，控制字段用来标识帧的顺序和重传行为，但由于该功能在 PPP 中并没有普遍实现，因此 PPP 帧中，其值固定为 0x03，表示无序号帧。

（4）协议字段：用于表示 PPP 帧封装的协议报文类型，其中，0x0021 代表 IP 报文，0x8021 代表 NCP 报文，0xC021 代表 LCP 报文，0xC023 代表 PAP 报文，0xC223 代表 CHAP 报文。

（5）信息字段：包含协议中指定协议的数据包，默认最大长度（不包含协议字段）称为最大接收单元（Maximum Receive Unit，MRU）（MRU 默认值为 1500 字节）。

4.2.3 PPP 的工作流程

PPP 的工作流程如图 4-11 所示。当用户拨号接入 ISP 后，就建立了一条用户计算机到 ISP 的物理连接。用户计算机向 ISP 发送一系列的 LCP 分组（封装成多个 PPP 帧），以便建立 LCP 连接。这些分组及其响应选择了将要使用的一些 PPP 参数。NCP 给新接入的用户计算机分配了一个临时的 IP 地址，这样用户计算机就成了 Internet 上一台有 IP 地址的主机。用户通信完毕后，NCP 释放网络层连接，收回原来释放出去的 IP 地址；LCP 释放数据链路层连接，通信双方释放物理连接。

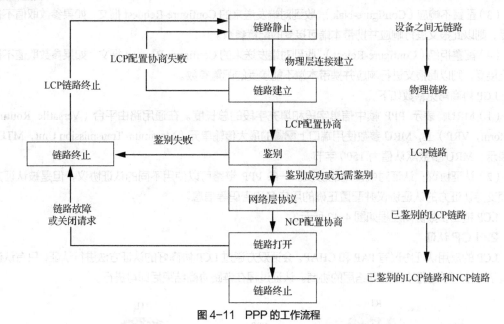

图 4-11 PPP 的工作流程

（1）链路不可用阶段（Dead 阶段）：表示物理层不可用。当通信双方均检测到物理线路激活时，会从链路不可用阶段变为链路建立阶段。

（2）链路建立阶段（Establish 阶段）：表示 LCP 参数协商中。其协商内容包括 MRU、认证方式等。协商成功后，会进入打开阶段，表示底层链路已经建立；反之，则返回到链路不可用阶段。

（3）验证阶段（Authentication 阶段）：该阶段不是必要阶段。如果需要验证，则在底层链路建立过程中进行验证。若验证通过或无验证，则进入网络层协议阶段；反之，则进入链路终止阶段，再返回到链路不可用阶段。

（4）网络层协议阶段（Network 阶段）：此阶段进行 NCP 协商。通过 NCP 协商来选择和配置一个网络层协议并进行网络参数的协商。只有相应的网络参数协商成功后，才会建立网络层通信；反之，则会进入链路终止阶段，再进入链路不可用阶段。

（5）打开阶段（Open 阶段）：当 NCP 协商成功后，PPP 链路将保持通信状态。

（6）链路终止阶段（Terminate 阶段）：此阶段所有资源都被释放，通信双方将回到链路不可用阶段。

4.2.4 LCP 概述

1. LCP 主要内容

LCP 是 PPP 的一个子集，用于在 PPP 会话建立过程中进行链路控制。LCP 可以自动检测链路环境，如是否存在环路，还可以协商链路参数，如最大数据包长度，也可以协商使用何种认证协议等。

LCP 报文类型有以下 4 种。

（1）配置请求（Configure-Request）：数据链路层协商过程中发送的第一个报文，该报文表示点到点的双方开始进行数据链路层参数的协商。

（2）配置响应（Configure-Ack）：收到对端发送来的 Configure-Request 报文，如果参数取值完全被接受，则以此报文进行响应。

（3）配置不响应（Configure-Nak）：收到对端发送来的 Configure-Request 报文，如果参数取值不被接受，则以此报文进行响应并携带本端可接受的配置参数。

（4）配置拒绝（Configure-Reject）：收到对端发送来的 Configure-Request 报文，如果参数取值不被完全接受，则以此报文进行响应并携带本端不能接受的配置参数。

LCP 协商常见参数如下。

（1）MRU：表示 PPP 帧中信息字段和填充字段的总长度。在通用路由平台（Versatile Routing Platform，VRP）上，MRU 参数使用端口上配置的最大传输单元（Maximum Transmission Unit，MTU）来表示。MRU 参数默认值为 1500 字节。

（2）认证协议：认证对端使用的协议。一条 PPP 链路可以使用不同的认证协议，但是被认证方必须支持认证方的认证协议并配置正确的用户名和密码等信息。

LCP 链路参数协商过程如图 4-12 所示。

2. LCP 认证

LCP 的常用认证协议有 PAP 和 CHAP。会话双方通过 LCP 协商好的认证方法进行认证，只有认证通过了，才可以进行下面的网络层的协商。认证过程在链路协商结束后即可进行。

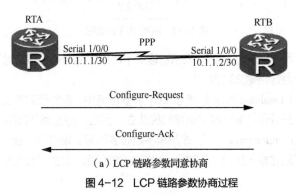

(a) LCP 链路参数同意协商

图 4-12 LCP 链路参数协商过程

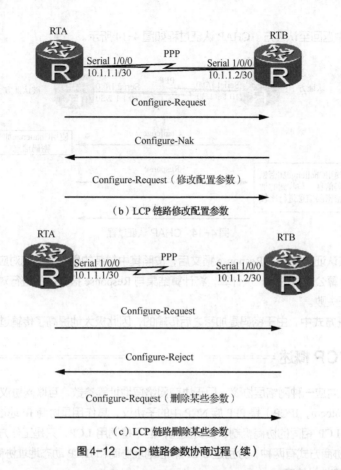

（b）LCP 链路修改配置参数

（c）LCP 链路删除某些参数

图 4-12　LCP 链路参数协商过程（续）

（1）PAP 认证：两次握手，明文传输密码，安全性低。被认证方将配置的用户名和密码信息使用 Authenticate-Request 报文以明文方式发送给认证方。认证方收到被认证方发送的用户名和密码之后，根据本地配置的用户名和密码数据库进行匹配，若匹配成功，则返回 Authenticate-Ack 报文，表示认证成功；否则返回 Authenticate-Nak 报文，表示认证失败。PAP 认证过程如图 4-13 所示。

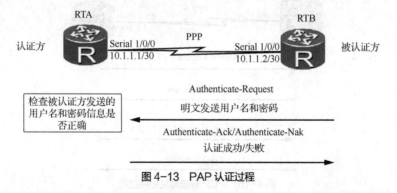

图 4-13　PAP 认证过程

（2）CHAP 认证：三次握手，密文传输密码。认证方发送一个 Challenge 报文给被认证方，报文中包含 Identifier 信息与一个随机产生的 Challenge 字符串，此 Identifier 即后续报文使用的 Identifier。被认证方收到该 Challenge 报文后，会进行一次加密计算，计算公式为 MD5（Identifier + 密码 + Challenge 字符串），从而得到一个 16 字节的摘要信息，然后将此信息和在端口上配置的 CHAP 用户名一起封装在 Response 报文中返回至认证方。CHAP 认证过程如图 4-14 所示。

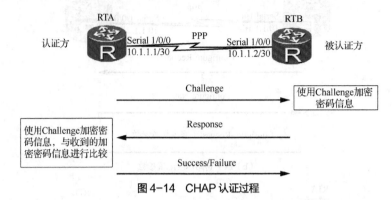

图 4-14　CHAP 认证过程

认证方收到被认证方发送的 Response 报文后，按照其中封装的用户名查找对应的密码，得到密码后，按照上面的计算公式再进行一次计算，将计算结果与 Response 报文中的进行对比，相同则认证成功，不相同则认证失败。

在 CHAP 认证方式中，由于密码是加密之后传输的，因此极大地提高了传输过程的安全性。

4.2.5　IPCP 概述

每一个 NCP 对应一种网络层协议，用于协商网络层地址等参数。互联网协议控制协议（Internet Protocol Control Protocol，IPCP）是 PPP 族 NCP 中的子协议，其作用是协商 IP 地址。

IPCP 使用与 LCP 相同的协商机制、报文类型，但不是调用 LCP，只是工作方式和报文类型等与 LCP 相同。IPCP 协商方式有两种，分别是 IPCP 静态地址协商和 IPCP 动态地址协商。

1. IPCP 静态地址协商

图 4-15 所示为 IPCP 静态地址协商过程，其中 RTA 为请求地址协商端（简称协商端），RTB 为地址分配被协商端（简称被协商端），具体过程如下。

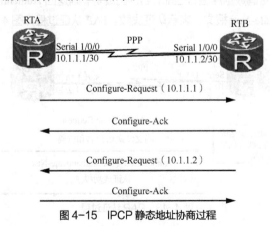

图 4-15　IPCP 静态地址协商过程

（1）协商端和被协商端都要发送包含本端的 IP 地址的 Configure-Request 报文给对端。

（2）协商端和被协商端都会收到包含对端的 IP 地址的 Configure-Request 报文，检查其中的 IP 地址，若 IP 地址是一个合法的单播 IP 地址且和本端的 IP 地址不冲突，则认为对端可以使用该 IP 地址，并且回应一个 Configure-Ack 报文。

2. IPCP 动态地址协商

图 4-16 所示为 IPCP 动态地址协商过程,其中 RTA 为请求地址协商端,RTB 为地址分配被协商端,具体过程如下。

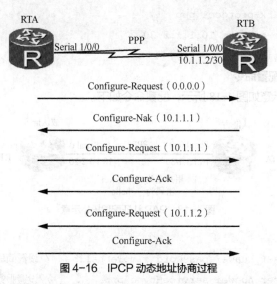

图 4-16 IPCP 动态地址协商过程

(1)协商端(没有 IP 地址)向被协商端发送一个 Configure-Request 报文,由于协商端没有 IP 地址,因此报文中包含的 IP 地址为 0.0.0.0,表示向被协商端请求 IP 地址。

(2)被协商端收到 Configure-Request 报文之后,认为其中的 IP 地址不合法,使用 Configure-Nak 回应一个新的 IP 地址(需提前建立 IP 地址池,并且与端口进行关联)。

(3)协商端收到 Configure-Nak 报文后,更新本端的 IP 地址,之后重新发送一个 Configure-Request 报文,此报文包含新的 IP 地址。

(4)被协商端收到新的 Configure-Request 报文后,认为其中的 IP 地址是合法的,回应一个 Configure-Ack 报文,并且会将包含本端 IP 地址的 Configure-Request 报文发送给协商端。

(5)协商端收到被协商端发送的 Configure-Request 报文后,认为其中的 IP 地址合法,会回应一个 Configure-Ack 报文。至此,IPCP 的协商完成。

4.2.6 PPP 的基础配置命令

PPP 配置拓扑示意如图 4-17 所示。

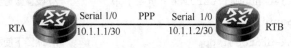

图 4-17 PPP 配置拓扑示意

1. 路由器端口编址

路由器端口编址的代码如下。

```
[RTA]interface Serial 1/0
[RTA-Serial1/0]link-protocol ppp  //端口下使能 PPP
[RTA-Serial1/0]ip address 10.1.1.1 30
```

```
[RTA-Serial1/0]quit

[RTB]interface Serial 1/0
[RTB-Serial1/0]link-protocol ppp
[RTB-Serial1/0]ip address 10.1.1.2 30
[RTB-Serial1/0]quit
```

2. PAP 认证方式配置命令

PAP 认证配置拓扑示意如图 4-18 所示，配置命令如下。

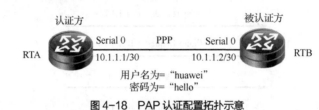

图 4-18　PAP 认证配置拓扑示意

```
[RTA]aaa
[RTA-aaa]local-user huawei password simple hello    //设置简单用户名和密码
[RTA-aaa]local-user huawei service-type ppp         //设置服务类型为 PPP
[RTA]interface Serial 0
[RTA-Serial0]link-protocol ppp                      //端口下使能 PPP
[RTA-Serial0]ppp authentication-mode pap            //设置认证方式为 PAP
[RTA-Serial0]ip address 10.1.1.1 30


[RTB]interface Serial 0
[RTB-Serial0]link-protocol ppp
[RTB-Serial0]ppp pap local-user huawei password simple hello   //对端设置 PAP 应用
[RTB-Serial0]ip address 10.1.1.2 30
```

3. CHAP 认证方式配置命令

如图 4-18 所示，修改 PAP 认证方式为 CHAP 认证方式，用户名和密码单独应用，配置命令如下。

```
[RTA]aaa
[RTA-aaa]local-user huawei password cipher hello    //创建用户名和密码
[RTA-aaa]local-user huawei service-type ppp         //设置服务类型为 PPP
[RTA]interface Serial 0
[RTA-Serial0]link-protocol ppp                      //端口下使能 PPP
[RTA-Serial0]ppp authentication-mode chap           //设置认证方式为 CHAP
[RTA-Serial0]ip address 10.1.1.1 30


[RTB]interface Serial 0
[RTB-Serial0]link-protocol ppp
[RTB-Serial0]ppp chap user huawei                   //用户名应用 CHAP
```

```
[RTB-Serial0]ppp chap password cipher hello      //密码应用CHAP
[RTB-Serial0]ip address 10.1.1.2 30
```

4. IPCP 动态协商 IP 地址配置命令

IPCP 动态协商 IP 地址配置拓扑示意如图 4-19 所示，路由器的 Serial 0 端口设置 PPP 方式，在 RTA 方启用动态获取 IP 地址命令，在 RTB 方设置 IP 地址提供命令，代码如下。

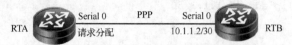

图 4-19　IPCP 动态协商 IP 地址配置拓扑示意

```
[RTA]interface Serial 0
[RTA-Serial0]link-protocol ppp
[RTA-Serial0]ip address ppp-negotiate   //动态获取IP地址


[RTB]interface Serial 0
[RTB-Serial0]link-protocol ppp
[RTB-Serial0]ip address 10.1.1.2 30
[RTB-Serial0]remote address 10.1.1.1    //提供IP地址
```

【任务实施】

实验 2　配置 PPP

实验目的：
（1）掌握 PPP 的原理。
（2）熟悉 PPP 的基本配置及地址借用。
（3）掌握 PPP 的 PAP 认证和 CHAP 认证配置方法。

V4-2　配置 PPP（PAP 认证）

V4-3　配置 PPP（CHAP 认证）

V4-4　配置 PPP（IPCP 动态协商 IP 地址）

1. 实验拓扑

PPP 配置拓扑示意如图 4-20 所示，要求用户与运营商之间完成 PPP 认证通信，即完成广域网串行链路 PPP 封装协议及 PPP 的 PAP 认证和 CHAP 认证配置。

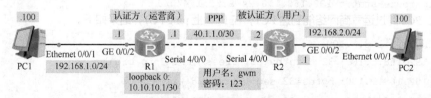

图 4-20　PPP 配置拓扑示意

2. 实验过程

（1）PC 编址，具体如下。
PC1：IP 地址为 192.168.1.100/24，网关为 192.168.1.1。

PC2：IP 地址为 192.168.2.100/24，网关为 192.168.2.1。

（2）路由器 IP 地址设置，代码如下。

```
[R1]int g0/0/2
[R1-GigabitEthernet0/0/2]ip add 192.168.1.1 24
[R1-GigabitEthernet0/0/2]int s4/0/0
[R1-Serial4/0/0]ip add 40.1.1.1 30

[R2]int g0/0/2
[R2-GigabitEthernet0/0/2]ip add 192.168.2.1 24
[R2-GigabitEthernet0/0/2]int s4/0/0
[R2-Serial4/0/0]ip add 40.1.1.2 30
```

（3）串口封装协议设置。华为设备串口封装协议默认为 PPP，所以不需要在串口上配置封装协议，代码如下。

```
[R1-Serial4/0/0]dis this
[V200R003C00]
#
interface Serial4/0/0
link-protocol ppp
 ip address 40.1.1.1 255.255.255.252
#
return
[R2-Serial4/0/0]dis this
[V200R003C00]
#
interface Serial4/0/0
link-protocol ppp
 ip address 40.1.1.2 255.255.255.0
#
Return
```

（4）配置静态路由，实现 PC1 与 PC2 的互通，代码如下。

```
[R1]ip route-static 192.168.2.0 24 Serial 4/0/0
[R2]ip route-static 192.168.1.0 24 Serial 4/0/0
```

此时，PC2 跨过运营商网络能到达 PC1，在 PC2 上 ping PC1，结果如下。

```
PC>ping 192.168.1.100
Ping 192.168.1.100: 32 data bytes, Press Ctrl_C to break
From 192.168.1.100: bytes=32 seq=1 ttl=126 time=16 ms
From 192.168.1.100: bytes=32 seq=2 ttl=126 time=15 ms
From 192.168.1.100: bytes=32 seq=3 ttl=126 time=17 ms
From 192.168.1.100: bytes=32 seq=4 ttl=126 time=31 ms
From 192.168.1.100: bytes=32 seq=5 ttl=126 time=16 ms
```

```
--- 192.168.1.100 ping statistics ---
5 packet(s) transmitted
5 packet(s) received
```

（5）PAP 认证。在认证方设置 PAP 认证，如设置用户名（gwm）和密码（123），在被认证方设置 PAP 连接测试，具体代码如下。

```
[R1]aaa
[R1-aaa]local-user gwm password cipher 123    //认证方设置用户名和密码分别为"gwm"和"123"
Info: Add a new user.
[R1-aaa]local-user gwm service-type ppp       //设置用户服务为 PPP 服务
[R1-aaa]int Serial 4/0/0
[R1-Serial4/0/0]ppp authentication-mode ?
  chap  Enable CHAP authentication
  pap   Enable PAP authentication
[R1-Serial4/0/0]ppp authentication-mode pap  //在端口上设置认证方式为 PAP
[R1-Serial4/0/0]dis this
[V200R003C00]
#
interface Serial4/0/0
 link-protocol ppp
 ppp authentication-mode pap
 ip address 40.1.1.1 255.255.255.252
#
return


[R2]int Serial 4/0/0
[R2-Serial4/0/0]ppp pap local-user gwm password simple 123
                //被认证方使用 PAP 认证，输入用户名和密码并进行登录
[R2-Serial4/0/0]dis this
[V200R003C00]
#
interface Serial4/0/0
 link-protocol ppp
 ppp pap local-user gwm password simple 123    //用户名和密码与认证方的配置信息一致
 ip address 40.1.1.2 255.255.255.252
#
return
[R2-Serial4/0/0]shutdown         //关闭端口
[R2-Serial4/0/0]undo  shutdown //启用端口
```

被认证方使用用户名"gwm"和密码"123"登录，若与认证方的配置信息一致，则 PAP 认证通过，LCP 启用，IPCP 启用，代码如下。

```
[R2]dis int Serial4/0/0
Serial4/0/0 current state : UP
Line protocol current state : UP
Last line protocol up time : 2021-11-11 20:10:19 UTC-08:00
Description:HUAWEI, AR Series, Serial4/0/0 Interface
Route Port, The Maximum Transmit Unit is 1500, Hold timer is 10(sec)
Internet Address is 40.1.1.2/30
Link layer protocol is PPP          //串行连接协议 PPP
LCP opened, IPCP opened             //LCP 启用，IPCP 启用
```

若被认证方使用其他用户名和密码登录，则 PAP 认证不能通过，LCP 关闭，代码如下。

```
[R2-Serial4/0/0]dis this
[V200R003C00]
#
interface Serial4/0/0
 link-protocol ppp
 ppp pap local-user hw password simple 123456    //用户名和密码与认证方的配置信息不一致
 ip address 40.1.1.2 255.255.255.252
#
return
[R2-Serial4/0/0]shutdown          //关闭端口
[R2-Serial4/0/0]undo  shutdown //启用端口
[R2]display int Serial 4/0/0
Serial4/0/0 current state : UP
Line protocol current state : DOWN
Description:HUAWEI, AR Series, Serial4/0/0 Interface
Route Port, The Maximum Transmit Unit is 1500, Hold timer is 10(sec)
Internet Address is 40.1.1.2/30
Link layer protocol is PPP
LCP closed                  //LCP 处于关闭状态
```

（6）CHAP 认证。把认证方设置的 PAP 认证删除，改为 CHAP 认证，仍然设置用户名和密码为 "gwm" 和 "123"，删除被认证方设置的 PAP 方式，改为 CHAP 方式，代码如下。

```
[R1]int Serial 4/0/0
[R1-Serial4/0/0]ppp authentication-mode chap  //设置 CHAP 认证
[R1-Serial4/0/0]dis this
[V200R003C00]
#
interface Serial4/0/0
 link-protocol ppp
 ppp authentication-mode chap
 ip address 40.1.1.1 255.255.255.252
```

```
#
return

[R2-Serial4/0/0]dis this
[V200R003C00]
#
interface Serial4/0/0
 link-protocol ppp
 ppp pap local-user hw password simple 123456
 ip address 40.1.1.2 255.255.255.252
#
return
[R2-Serial4/0/0]undo ppp pap local-user          //删除 PAP 方式
[R2-Serial4/0/0]ppp chap user gwm                //设置以 CHAP 方式认证用户名
[R2-Serial4/0/0]ppp chap password simple 123     //设置以 CHAP 方式认证密码
[R2-Serial4/0/0]dis this
[V200R003C00]
#
interface Serial4/0/0
 link-protocol ppp
 ppp chap user gwm
 ppp chap password simple 123
 ip address 40.1.1.2 255.255.255.252
#
return
[R2-Serial4/0/0]shutdown               //关闭端口
[R2-Serial4/0/0]undo shutdown          //启用端口
```

若被认证方使用用户名"gwm"和密码"123"登录，与认证方的配置信息一致，则 CHAP 认证通过，LCP 启用，IPCP 启用。

（7）IPCP 动态协商 IP 地址。被认证方假设没有静态配置 IP 地址，即 Serial 4/0/0 端口没有 IP 地址，把前面静态配置的 IP 地址 40.1.1.2/30 删除，那么需要向认证方申请一个 IP 地址。这里假设认证方通过认证，动态分配了一个 IP 地址 40.1.1.2 给认证方 Serial 4/0/0 端口，具体代码如下。

```
[R2]int Serial 4/0/0
[R2-Serial4/0/0]shutdown
[R2-Serial4/0/0]undo ip add 40.1.1.2 30
[R2-Serial4/0/0]dis ip int brief
Interface                IP Address/Mask      Physical     Protocol
GigabitEthernet0/0/0     unassigned           down         down
GigabitEthernet0/0/1     unassigned           down         down
GigabitEthernet0/0/2     192.168.2.1/24       up           up
```

```
NULL0                           unassigned            up         up(s)
Serial4/0/0                     unassigned            up         up
Serial4/0/1                     unassigned            down       down
```
[R2-Serial4/0/0]ip add ppp-negotiate //设置IPCP协商获取IP地址
[R2-Serial4/0/0]dis this
[V200R003C00]
#
interface Serial4/0/0
 link-protocol ppp
 ppp chap user gwm
 ppp chap password simple 123
 ip address ppp-negotiate
#
return

[R1]int Serial 4/0/0
[R1-Serial4/0/0]remote address 40.1.1.2   //分配IP地址40.1.1.2
[R1-Serial4/0/0]dis this
[V200R003C00]
#
interface Serial4/0/0
 link-protocol ppp
 ppp authentication-mode chap
 remote address 40.1.1.2
 ip address 40.1.1.1 255.255.255.252
#
return

[R2-Serial4/0/0]undo shutdown   //启用端口
[R2-Serial4/0/0]dis ip int brief   //查看获取IP地址的情况
```
Interface                  IP Address/Mask       Physical     Protocol
GigabitEthernet0/0/0       unassigned            down         down
GigabitEthernet0/0/1       unassigned            down         down
GigabitEthernet0/0/2       192.168.2.1/24        up           up
NULL0                      unassigned            up           up(s)
Serial4/0/0                40.1.1.2/32           up           up
Serial4/0/1                unassigned            down         down
```

【任务拓展】

本实验模拟广域网路由器互联，拓扑示意如图4-21所示。

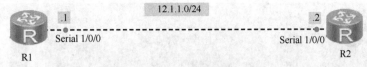

图 4-21　PPP 封装拓展拓扑示意

完成以下配置需求。

（1）配置设备名称和 IP 地址，R1 与 R2 使用 PPP 互联，R1 的 IP 地址手动配置，R2 的 IP 地址使用 PPP 由 R1 分配。

（2）R1 为认证方，R2 为被认证方，使用 CHAP 认证，并满足以下需求：R1 上使用 AAA 管理 [AAA 管理是一种网络管理架构，即 Authentication（身份验证）、Authorization（授权）和 Accounting（计费）]用户名和密码，用户名为 "huawei"，密码为 "hw123"，仅允许服务的类型为 PPP。

模块 5
网络安全技术

05

随着互联网相关技术的发展，网络的普及范围越来越广，上网成为人们生活中必不可少的一部分。因此，网络安全也关系到人们的日常学习、生活。特别是企业或机构不同办公地点之间数据的传输要经过互联网，因此更要保护数据传输的安全。例如，在校园网这一特殊的区域，保障网络内容的绿色、安全是首要条件。通过对校园内计算机访问地址进行限制可以保护学生在校园网中的联网安全，而 ACL 技术能帮助实现访问限制。

目前的互联网普遍使用 TCP/IP 进行基础通信，而 TCP/IP 本身并没有提供安全机制，这就导致数据在互联网传输的过程中容易被伪造、篡改或者窥视。VPN 是对企业内部网络的扩展，通过 VPN 可以帮助远程用户、企业或单位分支机构的内部网络建立可信的安全连接，并保证数据的安全传输。IPSec VPN 可以有效地保护 IP 数据报的安全，它提供了一套健壮的安全机制，可为 IP 上层协议提供安全保障。本模块主要介绍网络安全技术，包括 ACL 的配置、利用 IPSec 的 VPN 技术进行配置等。

【学习目标】

知识目标
（1）掌握 ACL 的定义、作用、分类。
（2）了解基本 ACL 和高级 ACL 的应用场景。
（3）理解 IPSec 封装模式。

技能目标
（1）掌握 ACL 在企业网络中的应用方法。
（2）掌握基本 ACL 与高级 ACL 的配置方法。
（3）掌握 ACL 在端口上的应用方法。
（4）掌握 VPN 中 IPSec 安全策略的设计过程。

素质目标
（1）通过讲解 ACL 的作用，提高学生的网络安全防护意识。
（2）培养学生的职业认知，教育学生养成良好的职业习惯。
（3）通过讲解我国华为网络安全产品的研发，以及在国内外市场的应用前景，培养学生的民族自豪感。

任务 5.1　ACL 的配置

【知识准备】

5.1.1　ACL 概述

在互联网发达的今天，网络服务器的数据访问压力越来越大，网络安全问题越发严峻。在中小型企业网络中，硬件式防火墙投入成本较高，而访问控制列表（Access Control List，ACL）规则灵活，只需要在路由器上定义相应参数设置，使数据包有选择地通过路由器，以达到过滤效果，起到软件式防火墙的作用。

ACL 是一种基于包过滤的流控制技术，它是由一系列 permit 语句和 deny 语句组成的（后面跟着条件列表）有序规则的列表，通过匹配报文的相关信息实现对报文的分类。ACL 规则灵活多变，设备可以定义一系列不同的 ACL 规则，以对不同类型的数据报文进行分类处理，从而控制网络的访问行为，防止网络攻击，提高网络性能，限制网络流量，最终保障网络传输的可靠性和稳定性。ACL 的主要作用是控制设备之间的数据包的互通，通过检查数据包的 3 层 IP 的头部信息和 4 层传输层协议的头部信息来有针对性地抓取数据包。针对 3 层 IP 的头部信息，ACL 可以抓取数据包中的源 IP 地址、目的 IP 地址、协议号等；针对 4 层传输层协议的头部信息，ACL 可以抓取数据包的源端口号、目的端口号等。ACL 主要具有以下 6 个特性。

① 具有包过滤防火墙功能。
② 提供网络地址转换（Network Address Translation，NAT）服务。
③ 具有服务质量（Quality of Service，QoS）的数据分类方式。
④ 具有路由策略和过滤功能。
⑤ 提供 IPSec VPN。
⑥ 按需拨号。

图 5-1 所示为 ACL 的包过滤技术，ACL 可以对进出的数据包逐个过滤后选择丢弃（拒绝）或允许通过（允许）。ACL 应用于端口上，每个端口的出方向和入方向均需进行过滤，仅当数据包经过一个端口时，才能被此端口的当前方向的 ACL 过滤。

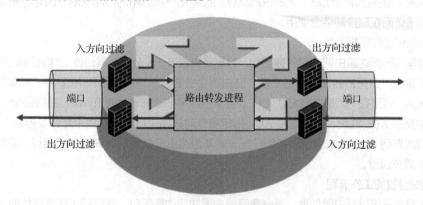

图 5-1　ACL 的包过滤技术

ACL 通常应用在企业的出口控制上，通过实施 ACL，可以有效地部署企业网络出网策略。随着局域网内部网络资源的增加，一些企业已经开始使用 ACL 来控制对局域网内部资源的访问能力，进而来保障网络资源的安全性。

5.1.2 ACL 的应用场景

ACL 的主要作用是控制设备之间的数据包的互通，其主要应用场景如下。

（1）ACL 应用在 Telnet 模块中，可以使设备作为 Telnet 服务器时，对以 Telnet 方式登录到设备的 Telnet 客户端加以控制，从而有效防止未经授权用户的非法接入。

（2）ACL 应用在 QoS 的流策略/简化流策略中，可以实现不同网段用户之间访问权限的限制，从而避免用户之间随意访问形成安全隐患。

（3）ACL 应用在 QoS 的流策略/简化流策略中，可以实现特定用户主机在特定时间范围内上网权限的限制。

（4）ACL 应用在 QoS 的流策略/简化流策略中，可以实现对不同流量进入网络的速率的监督，对超出部分的流量进行"惩罚"，使进入的流量被限制在一个合理的范围之内，从而保护网络资源和用户利益。

（5）ACL 可以应用在各种动态路由协议中，对路由协议发布和接收的路由信息进行过滤。

（6）ACL 应用在简单网络管理协议（Simple Network Management Protocol，SNMP）模块中，可以使网络管理员对设备的管理权限得到控制，从而有效防止非法网络管理员对设备进行操作。

（7）ACL 可以应用在 NAT 过滤功能中，让 NAT 设备对外网发送到内网的流量进行过滤。

（8）ACL 应用在内外网络边缘处，用来防止外部网络对内部网络的入侵，也可以用来保护网络内部大型机和重要的资源（如数据）。

5.1.3 ACL 的包过滤流程

ACL 增强了在路由器端口上控制数据包出入的灵活性，可以限制网络流量，也可以限制用户和设备对网络的使用权限。一般的路由器都提供了 ACL 功能，对于一些网络流量不是很大的场景，完全可以借助 ACL 实现防火墙的部分功能，但它不能完全代替防火墙，因为 ACL 不能像专业的防火墙那样对相应数据包进行分析。因此 ACL 只可用于网络的初级防范。不仅如此，ACL 过滤的依据只是数据包第 3 层和第 4 层头部部分信息，无法识别具体的控制用户和应用的内部权限。因此，ACL 要和系统级和应用级的访问权限控制结合使用。

1. 入站包过滤工作流程

入站包是指已到达路由器端口的数据包，将被路由器进一步处理，其过滤工作流程如图 5-2 所示。

在数据包通过某路由器进入网络之前，路由器会检查该入站口是否在 ACL 中，如果不在，则直接从该口进入；如果在，则路由器将依照从上到下的顺序依次将该数据和 ACL 进行匹配，逐条执行。如果与其中某条 ACL 匹配，则根据该 ACL 指定操作对数据进行相应处理（允许或拒绝），并停止继续查询；如果直到 ACL 的末尾也未找到匹配，则华为路由设备默认允许数据包通过，思科路由设备默认拒绝数据包通过。

2. 出站包过滤工作流程

出站包是指已经过路由器处理，准备离开路由器端口的数据包，其过滤工作流程如图 5-3 所示。

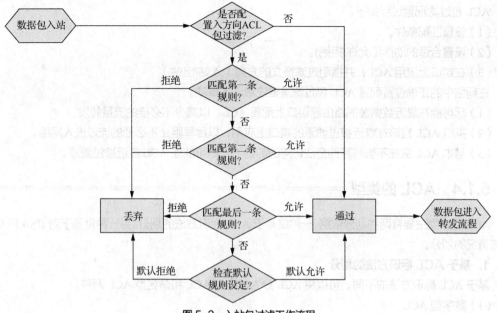

图 5-2 入站包过滤工作流程

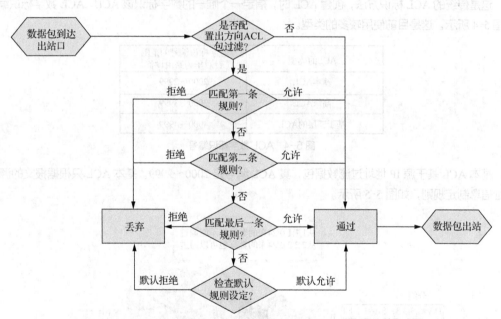

图 5-3 出站包过滤工作流程

无论路由器上有没有 ACL，接到数据包后，当数据进入某个入站口时，路由器首先对其进行检查，看其是否可路由，不可路由则丢弃；否则通过查询路由表发现该路由的详细信息及对应的出站口。假设可路由，则找出要将其送出站的端口，此时路由器检查该出站口有没有被编入 ACL，没有则直接从该口送出；如果有 ACL，则路由器将依照从上到下的顺序依次将该数据和 ACL 进行匹配，逐条执行。如果与其中某条 ACL 匹配，则根据该 ACL 指定操作对数据进行相应处理（允许或拒绝），并停止继续查询；如果直到 ACL 的末尾也未找到匹配，则华为路由设备默认允许数据包通过，思科路由设备默认拒绝数据包通过。

ACL 包过滤配置过程如下。
（1）设置匹配条件。
（2）设置合适的动作（允许/拒绝）。
（3）在端口上应用 ACL，并指明过滤报文的方向（入站/出站）。
在网络中的正确位置配置 ACL 包过滤需要注意以下事项。
（1）尽可能在靠近数据源的路由器端口上配置 ACL，以减少不必要的流量转发。
（2）高级 ACL 应该在靠近被过滤源的端口上应用，以尽早阻止不必要的流量进入网络。
（3）基本 ACL 应在不影响其他合法访问的前提下，尽可能使 ACL 靠近被过滤源。

5.1.4　ACL 的类型

ACL 的类型主要有两种划分依据，一种是基于 ACL 标识方法的划分，另一种是基于对 IPv4 和 IPv6 支持情况的划分。

1. 基于 ACL 标识方法的划分

基于 ACL 标识方法的不同，可以将 ACL 分为数字型 ACL 和命名型 ACL 两种。

（1）数字型 ACL

这是传统的 ACL 标识方法，创建 ACL 时，指定一个唯一的数字标识该 ACL。ACL 数字标识编号如图 5-4 所示。这是目前使用较多的类型。

ACL的类型	数字标识编号的范围 （针对华为路由设备）
基本ACL	2000～2999
高级ACL	3000～3999
基于二层的ACL	4000～4999

图 5-4　ACL 数字标识编号

基本 ACL 基于源 IP 地址过滤数据包，其 ACL 编号是 2000～2999。基本 ACL 只根据报文的源 IP 地址信息制定规则，如图 5-5 所示。

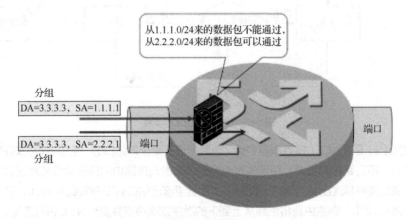

图 5-5　基本 ACL

高级 ACL 基于源 IP 地址、目的 IP 地址、指定协议、端口和标志来过滤数据包，其 ACL 编号是 3000～3999，如图 5-6 所示。

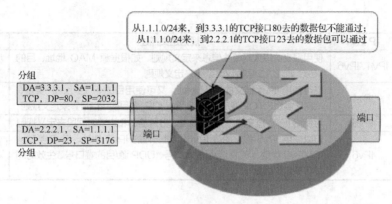

图 5-6 高级 ACL

基于二层的 ACL 根据报文的源 MAC 地址、目的 MAC 地址、IEEE 802.1p 优先级、二层协议类型等二层信息制定匹配规则，其 ACL 编号是 4000~4999，如图 5-7 所示。

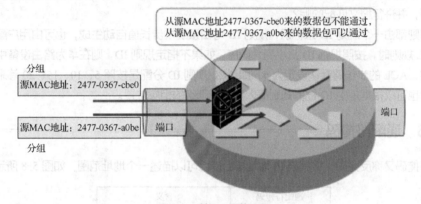

图 5-7 基于二层的 ACL

（2）命名型 ACL

用户在创建 ACL 时可以为其指定编号，不同的编号对应不同类型的 ACL。同时，为了便于记忆和识别，用户还可以创建命名型 ACL，即在创建 ACL 时为其设置名称。命名型 ACL 可以是"名称 数字"的形式，即在定义命名型 ACL 的同时指定 ACL 编号。如果不指定编号，则系统会自动为其分配一个数字型 ACL 的编号。

命名型 ACL 一旦创建成功，便不允许用户再修改其名称。如果删除 ACL 名称，则表示删除整个 ACL。仅基本 ACL 与基本 ACL6，以及高级 ACL 与高级 ACL6 可以使用相同的 ACL 名称；其他类型 ACL 不能使用相同的 ACL 名称。

2. 基于对 IPv4 和 IPv6 支持情况的划分

基于对 IPv4 和 IPv6 支持情况的 ACL 分类如表 5-1 所示。

表 5-1 基于对 IPv4 和 IPv6 支持情况的 ACL 分类

分类	IP 版本	规则定义描述	编号范围
基本 ACL	IPv4	仅使用报文的源 IP 地址、分片信息和生效时间段来定义规则	2000~2999
高级 ACL	IPv4	既可使用 IPv4 报文的源 IP 地址，又可使用目的 IP 地址、IP 类型、ICMP 类型、TCP 源/目的端口号、UDP 源/目的端口号、生效时间段等来定义规则	3000~3999

续表

分类	IP 版本	规则定义描述	编号范围
基于二层的 ACL	IPv4、IPv6	使用报文的以太网帧头信息来定义规则，如根据源 MAC 地址、目的 MAC 地址、二层协议类型等来定义规则	4000~4999
用户自定义的 ACL	IPv4	既可使用 IPv4 报文的源 IP 地址，又可使用目的 IP 地址、IP 类型、ICMP 类型、TCP 源/目的端口号、UDP 源/目的端口号等来定义规则	6000~6031
基本 ACL6	IPv6	可使用 IPv6 报文的源 IPv6 地址、分片信息和生效时间段来定义规则	2000~2999
高级 ACL6	IPv6	可以使用 IPv6 报文的源 IPv6 地址、目的 IPv6 地址、IPv6 类型、ICMPv6 类型、TCP 源/目的端口号、UDP 源/目的端口号、生效时间段等来定义规则	3000~3999

5.1.5 ACL 步长

ACL 中的每条规则都有自己的编号，这个编号在该 ACL 中是唯一的。在创建规则时，可以人工为其指定一个编号，如未人工指定编号，则由系统自动为其分配一个编号。系统是根据什么样的顺序来选择规则，并进行报文匹配的呢？

每条规则都由一个规则 ID 来标识，规则 ID 可由系统按步长值自动生成，也可由用户配置。在定义一个 ACL 规则时，按照规则 ID 从小到大排序。如果不指定规则 ID，则在华为路由设备中由默认的步长值设定，ACL 的步长值默认设定为 5，即 ACL 规则 ID 分配是按照 5、10、15、20 等来分配的。当然，用户也可以根据规则 ID 把新规则插入某一规则组的相应位置。

5.1.6 通配符掩码

通配符掩码又称反掩码，将其与 IP 地址结合使用可以描述一个地址范围，如图 5-8 所示。

通配符掩码	含义
0.0.0.255	只比较前24位
0.0.3.255	只比较前22位
0.255.255.255	只比较前8位

图 5-8 通配符掩码

通配符掩码和子网掩码相似，但含义不同，0 表示对应位需匹配，1 表示对应位可不进行匹配（即无所谓匹配与否），如图 5-9 所示。

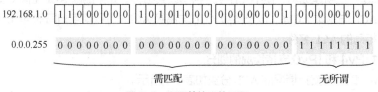

图 5-9 通配符掩码的匹配

特殊的通配符掩码有以下两种形式。

（1）精确匹配某 IP 地址。例如，精确匹配 192.168.1.1，如 192.168.1.1 0.0.0.0=192.168.1.1 0。

（2）匹配所有 IP 地址。例如，0.0.0.0 255.255.255.255=any。

例如，匹配 192.168.1.0/24 这个子网中最后一个 8 位组为奇数的 IP 地址，有 192.168.1.1、192.168.1.3、

192.168.1.5 等，如图 5-10 所示。

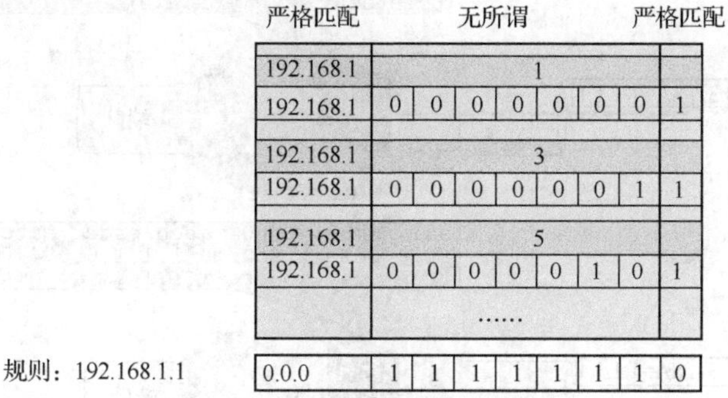

	严格匹配				无所谓				严格匹配
192.168.1					1				
192.168.1	0	0	0	0	0	0	0	0	1
192.168.1					3				
192.168.1	0	0	0	0	0	0	0	1	1
192.168.1					5				
192.168.1	0	0	0	0	0	0	1	0	1
					……				
规则：192.168.1.1	0.0.0	1	1	1	1	1	1	1	0

图 5-10 具体示例

5.1.7 ACL 的配置命令和匹配顺序

1. 配置 ACL

使用 acl [number]命令可指定 ACL 编号，用来创建一个 ACL，并进入 ACL 视图。示例如下。

```
[sysname]acl 2000
```

2. 定义规则

（1）基本 ACL：制定要匹配的源 IP 地址范围，指定动作是 permit 或 deny。基本格式如下。

```
[sysname-acl-basic-2000] rule [ rule-id ] { deny | permit } source{ sour-addr sour-wildcard | any } | time-range time-name ]
```

其中，rule 可以增加或修改基本 ACL 的规则；deny 指定拒绝符合条件的数据包；permit 指定允许符合条件的数据包；source 指定 ACL 规则匹配报文的源地址信息；any 表示任意源地址。

（2）高级 ACL：需要配置规则来匹配源 IP 地址、目的 IP 地址、IP 承载的协议类型、协议端口号等信息，指定动作是 permit 或 deny。基本格式如下。

```
[sysname-acl-adv-3000] rule [ rule-id ] { deny | permit } protocol [ destination { dest-addr dest-wildcard | any } | destination-portoperator port1 [ port2 ] established | fragment | source{ sour-addr sour-wildcard | any } | source-portoperator port1 [ port2 ] | time-range time-name ]
```

（3）基于二层的 ACL：需要配置规则来匹配源 MAC 地址、目的 MAC 地址、IEEE 802.1p 优先级、二层协议类型等二层信息，指定动作是 permit 或 deny。基本格式如下。

```
[sysname-acl-ethernetframe-4000] rule [ rule-id ] { deny | permit } [ cos VLAN-pri | dest-mac dest-addr dest-mask | lsap lsap-code lsap-wildcard | source-mac source-addr source-mask | time-range time-name ]
```

3. 应用 ACL

在端口上应用 ACL 时，需要使用 traffic-filter 语句才能使配置的 ACL 包过滤生效。注意，要指明在端口上应用的方向是 outbound 还是 inbound。基本格式如下。

```
[sysname-Serial2/0 ] traffic-filter { inbound | outbound }{ acl-number |name acl-name }
```

4. ACL 的匹配顺序

匹配顺序指 ACL 中规则的优先级，不同匹配顺序将导致不同的结果，如图 5-11 所示。

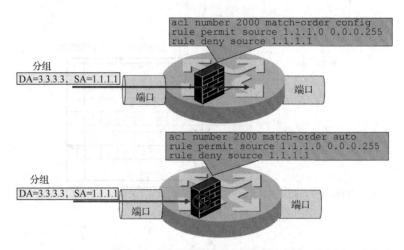

图 5-11 不同匹配顺序将导致不同的结果

ACL 支持两种匹配顺序,分别是配置顺序(config)和自动排序(auto)。
(1)配置顺序:按照用户配置规则的先后顺序进行规则匹配。
(2)自动排序:按照深度优先的顺序进行规则匹配,即地址范围小的规则被优先进行匹配。

【任务实施】

实验 1　配置 ACL

实验目的:
(1)掌握基本 ACL 的配置与调用方法。
(2)熟悉高级 ACL 的配置与调用方法。

1. 基本 ACL 的配置与调用

根据图 5-12 所示的拓扑示意完成基本 ACL 的配置与调用实验。这里要求 PC1 不能访问 PC2 和 PC3,但是可以访问 R1。

V5-1　基本 ACL 的配置与调用

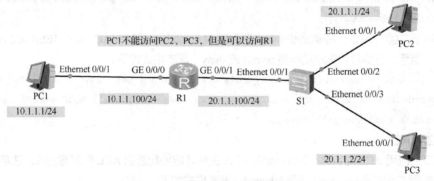

图 5-12　基本 ACL 实验配置拓扑示意

(1)设置路由器端口 IP 地址,代码如下。

```
[R1]int g0/0/0
[R1-GigabitEthernet0/0/0]ip address 10.1.1.100 24
```

```
[R1-GigabitEthernet0/0/0]int g0/0/1
[R1-GigabitEthernet0/0/1]ip address 20.1.1.100 24
```

这里省略配置 PC1、PC2、PC3 的 IP 地址的过程,但请注意 PC1、PC2、PC3 的网关分别设置为 10.1.1.100、20.1.1.100、20.1.1.100。

(2)配置基本 ACL,代码如下。

```
[R1]acl 2000
[R1-acl-basic-2000]rule ?
  INTEGER<0-4294967294>  ID of ACL rule
  deny                   Specify matched packet deny
  permit                 Specify matched packet permit
[R1-acl-basic-2000]rule 5 ?
  deny         Specify matched packet deny
  description  Specify rule description
  permit       Specify matched packet permit
[R1-acl-basic-2000]rule 5 deny source 10.1.1.1 ?
  IP_ADDR<X.X.X.X>  Wildcard of source
  0                 Wildcard bits : 0.0.0.0 ( a host )
[R1-acl-basic-2000]rule 5 deny source 10.1.1.1 0.0.0.0
[R1-acl-basic-2000]dis this
[V200R003C00]
#
acl number 2000
 rule 5 deny source 10.1.1.1 0
#
return
```

(3)调用基本 ACL。根据实验要求,允许 PC1 的流量通过 R1,但不能传送给 PC2 和 PC3,需在 R1 的 GE 0/0/1 端口处应用定义的 ACL,代码如下。

```
[R1-acl-basic-2000]int g0/0/1
[R1-GigabitEthernet0/0/1]traffic-filter ?
  inbound   Apply ACL to the inbound direction of the interface
  outbound  Apply ACL to the outbound direction of the interface
[R1-GigabitEthernet0/0/1]traffic-filter outbound ?
  acl   Specify ACL to match
  ipv6  Specify IPv6
[R1-GigabitEthernet0/0/1]traffic-filter outbound acl ?
  INTEGER<2000-2999>  Basic access-list
  INTEGER<3000-3999>  Advanced access-list
  INTEGER<4000-4999>  L2 access-list
  name                Specify a named ACL
[R1-GigabitEthernet0/0/1]traffic-filter outbound acl 2000
[R1-GigabitEthernet0/0/1]
```

（4）验证 ACL。在 PC1 上使用 ping 命令，分别测试到 R1 的 GE 0/0/1、PC2、PC3 的连通性，结果如图 5-13 所示。

图 5-13 基本 ACL 测试结果

2. 高级 ACL 的配置与调用

根据图 5-14 所示的拓扑示意完成高级 ACL 的配置与调用实验。这里要求 PC1 不能访问 PC2，但是可以访问 R1 和 PC3。

V5-2 高级 ACL 的配置与调用

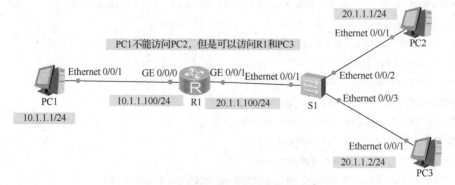

图 5-14 高级 ACL 实验配置拓扑示意

（1）配置 PC 端 IP 地址（步骤略）。

（2）配置高级 ACL，代码如下。

```
[R1]acl 3000
[R1-acl-adv-3000]rule deny ip source 10.1.1.1 0 destination 20.1.1.1 0
[R1-acl-adv-3000]quit
```

（3）调用高级 ACL，代码如下。

```
[R1]int g0/0/0
[R1-GigabitEthernet0/0/0]traffic-filter inbound acl 3000
[R1-GigabitEthernet0/0/0]quit
```

需要注意的是，由于在基本 ACL 的配置与调用实验中，已经在 R1 的 GE 0/0/1 端口上应用了 ACL 2000，因此为了保证高级 ACL 配置实验的效果，需要先清除 GE 0/0/1 端口上的 ACL 应用，具体代码如下。

```
[R1]int g0/0/1
[R1-GigabitEthernet0/0/1]dis this
[V200R003C00]
#
interface GigabitEthernet0/0/1
 ip address 20.1.1.100 255.255.255.0
 traffic-filter outbound acl 2000
#
return
[R1-GigabitEthernet0/0/1]undo traffic-filter out
[R1-GigabitEthernet0/0/1]undo traffic-filter outbound ?
  ipv6  Specify IPv6
<cr>  Please press ENTER to execute command
[R1-GigabitEthernet0/0/1]undo traffic-filter outbound
```

（4）测试高级 ACL，结果如下。

```
pc>ping 20.1.1.1
正在 Ping 20.1.1.1 具有 32 字节的数据：
来自 20.1.1.1 的回复：无法访问目标主机。
来自 20.1.1.1 的回复：无法访问目标主机。
来自 20.1.1.1 的回复：无法访问目标主机。
来自 20.1.1.1 的回复：无法访问目标主机。
20.1.1.1 的 Ping 统计信息：
数据包：已发送 = 4，已接收 = 0，丢失 = 4 (100% 丢失)


pc>ping 20.1.1.2
正在 Ping 20.1.1.2 具有 32 字节的数据：
来自 20.1.1.2 的回复：字节=32 时间<1ms TTL=64
```

来自 20.1.1.2 的回复：字节=32 时间<1ms TTL=64
来自 20.1.1.2 的回复：字节=32 时间<1ms TTL=64
来自 20.1.1.2 的回复：字节=32 时间<1ms TTL=64
20.1.1.2 的 Ping 统计信息：
    数据包：已发送 = 4，已接收 = 4，丢失 = 0 (0% 丢失)，
往返行程的估计时间（以毫秒为单位）：
最短 = 0ms，最长 = 0ms，平均 = 0ms

```
pc>ping 20.1.1.100
```
正在 Ping 20.1.1.100 具有 32 字节的数据：
来自 20.1.1.100 的回复：字节=32 时间<1ms TTL=64
来自 20.1.1.100 的回复：字节=32 时间<1ms TTL=64
来自 20.1.1.100 的回复：字节=32 时间<1ms TTL=64
来自 20.1.1.100 的回复：字节=32 时间<1ms TTL=64
20.1.1.100 的 Ping 统计信息：
    数据包：已发送 = 4，已接收 = 4，丢失 = 0 (0% 丢失)，
往返行程的估计时间（以毫秒为单位）：
最短 = 0ms，最长 = 0ms，平均 = 0ms

【任务拓展】

1. 基本 ACL 配置实验

根据图 5-15 所示的拓扑示意，完成基本 ACL 配置实验。

V5-3 任务 5.1
拓展任务 1 提示

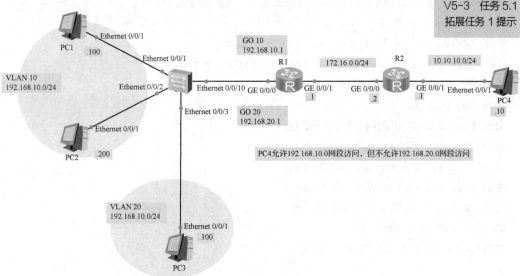

图 5-15 基本 ACL 配置实验拓展拓扑示意

PC（PC1、PC2、PC3 和 PC4）间的通信使用单臂路由实现，PC4 允许 192.168.10.0 网段访问，但不允许 192.168.20.0 网段访问。

2. 高级 ACL 配置实验

根据图 5-16 所示的拓扑示意，完成高级 ACL 配置实验。

PC 间通信使用单臂路由实现，VLAN 10 用户可以通过 HTTP 访问 10.10.10.100 服务器，但不能通过 HTTP 访问 10.10.10.50 服务器；VLAN 20 用户可以通过 HTTP 访问 10.10.10.50 服务器，但不能通过 HTTP 访问 10.10.10.100 服务器；所有设备都不能通过 FTP 访问 10.10.10.50 服务器，但所有设备间可以相互 ping 通。

V5-4　任务 5.1 拓展任务 2 提示

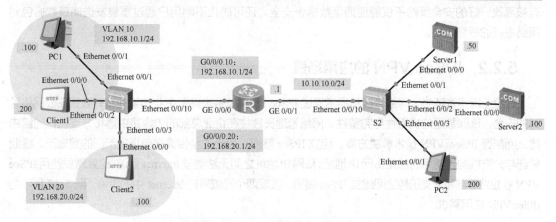

图 5-16　高级 ACL 配置实验拓展拓扑示意

任务 5.2　利用 IPSec 的 VPN 技术进行配置

【知识准备】

5.2.1　VPN 和 IPSec VPN 概述

1. VPN 技术

虚拟专用网（Virtual Private Network，VPN）技术就是在 Internet 中建立的连接两个或多个局域网的隧道。通过配置两端的路由设备，可以为两个局域网创建一条隧道，让两个局域网之间能够互相通信，通过加密和身份验证技术实现数据通信的安全，达到像专线一样的效果。

2. IPSec 技术

互联网安全协议（Internet Protocol Security，IPSec）是因特网工程任务组（Internet Engineering Task Force，IETF）定义的一个协议组。它不具体指某个协议，而是定义了一个框架，该框架涉及加密算法、验证算法、封装协议、封装模式、密钥有效期等。IPSec 对 IP 的分组进行加密和认证，通信双方在 IP 层通过加密、完整性校验、数据源认证等方式，来保证 IP 分组传输的数据报文在网络中的机密性、完整性和防重放。

（1）机密性：指对数据进行加密保护，用密文的形式传送数据。

（2）完整性：指对接收的数据进行认证，以判定报文是否被篡改。

（3）防重放：指防止恶意用户通过重复发送捕获的数据包所进行的攻击，即接收方会拒绝旧的或重复的数据包。

3. IPSec VPN

IPSec VPN 建立的前提如下：要想在两个站点之间安全地传输 IP 数据流，站点之间必须先进行协商，协商内容包括采用的加密算法、封装技术和密钥等。

随着企业网络发展规模的扩大，人们对网络安全技术的重视日益提升，安全策略问题受到广泛重视。IPSec 安全策略的引入，使传统 TCP/IP 网络数据通信的安全性得到了保障，在发送端可以用密文的形式对数据进行加密保护，在接收端对数据进行认证检测，以确保数据报文在传输过程中没有被篡改。好的安全策略不仅能使通信数据更安全，还可防止不明用户通过重复发送捕获数据包对网络进行的恶意攻击。

5.2.2 IPSec VPN 的应用场景

在企业网络管理中，网络管理员通常通过 IPSec VPN 技术来确保企业的某些私有数据在公网传输时的安全，确保数据的机密性与完整性。网络管理员往往在企业总部的边缘路由器和分支机构的路由器之间部署 IPSec VPN 技术解决方案，建立 IPSec 隧道，用于安全传输来自指定部门的数据流。企业总部与分支机构内使用的都是私网 IP 地址，私网 IP 地址之间无法通过 Internet 互访，这就需要使用 IPSec VPN 在企业总部和分支机构之间建立 IPSec 隧道，实现两个局域网在 Internet 中的传输。图 5-17 所示为 IPSec VPN 应用需求。

图 5-17　IPSec VPN 应用需求

IPSec VPN 的应用场景分为以下 3 种。

（1）站点到站点或网关到网关（Site-to-Site）：如 3 个机构分布在互联网的 3 个不同的地方，各使用一个网关相互建立 VPN 隧道，企业内网（若干 PC）之间的数据通过这些网关建立的 IPSec 隧道实现安全互联。

（2）端到端或 PC 到 PC（End-to-End）：两个 PC 之间的通信由两个 PC 之间的 IPSec 会话保护，而不是网关。

（3）端到站点或 PC 到网关（End-to-Site）：两个 PC 之间的通信由网关和异地 PC 之间的 IPSec 进行保护。

5.2.3 IPSec 架构

IPSec 位于传输层与网络层之间，它可以对传输层与应用层的数据进行保护。IPSec 不是一个单独的协议，IPSec 主要由身份认证头（Authentication Header，AH）协议、封装安全有效负载（Encapsulating Security Payload，ESP）协议和互联网密钥交换（Internet Key Exchange，IKE）协议套件组成，它通过 AH 协议和 ESP 协议来实现 IP 数据包的安全传送，IKE 协议提供密钥协商，建立和维护安全联盟（Security Association，SA）等服务，其架构如图 5-18 所示。

（1）AH 协议：提供数据源验证、数据完整性的校验、防报文重放等安全特性。AH 协议常用摘要算法（单向 Hash 函数）MD5 和 SHA-1 实现这些特性。

（2）ESP 协议：提供数据完整性确认、数据加密、防重放等安全特性，主要用来对 IP 报文进行加密，这样 IP 报文在传送的过程中能够保证在一个安全的网络中通信。ESP 协议通常使用 DES、3DES、AES 等加密算法实现数据加密，使用 MD5 或 SHA-1 实现数据完整性。

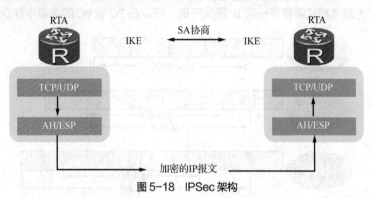

图 5-18　IPSec 架构

（3）IKE 协议：主要用于自动协商 AH 和 ESP 协议所使用的密码算法，建立和维护 SA 等服务。

VPN 只是 IPSec 的一种应用方式，它的目的是为 IP 提供高安全性，VPN 则是在实现这种安全特性的方式下产生的解决方案。

5.2.4　IPSec 的封装模式

IPSec 主要提供两种封装模式：传输（Transport）模式和隧道（Tunnel）模式。

1. 传输模式

在传输模式下，在 AH、ESP 处理前后 IP 报头保持不变，主要用于 End-to-End 应用场景中 PC 到 PC 的情况。因为公网不允许使用私网地址，所以不可部署在网关与网关之间。Windows 操作系统可部署传输模式，在传输模式下，AH 报头或 ESP 报头位于 IP 报头和传输层报头之间。传输模式下的报文封装格式如图 5-19 所示。

（1）传输模式的 AH：在 IP 报头插入 AH 报头，对整个数据包进行完整性校验。

（2）传输模式的 ESP：在 IP 报头插入 ESP 报头，在数据段后插入尾部及认证字段，对高层数据和 ESP 尾部进行加密，对 IP 数据包中的 ESP 报头、高层数据和 ESP 尾部进行完整性校验。

（3）传输模式的 AH-ESP：在 IP 报头之后插入 AH 和 ESP 报头，在数据段后插入尾部及认证字段，对高层数据和 ESP 尾部进行加密，对整个 IP 数据包进行完整性校验。

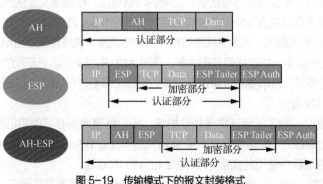

图 5-19　传输模式下的报文封装格式

2. 隧道模式

在隧道模式下，AH 或 ESP 报头封装在原始 IP 报头之前，并另外生成一个新的 IP 报头封装到 AH 或 ESP 之前。隧道模式下的报文封装格式如图 5-20 所示。隧道模式可以完全地对原始 IP 数据报进行认证和加密，而且可以使用 IPSec 对等体的 IP 地址来隐藏客户端的 IP 地址，主要用于 Site-to-Site 的应用场景。因为隧道模式需要多一层 IP 报头开销，所以在 PC 到 PC 的场景中建议使用传输模式。

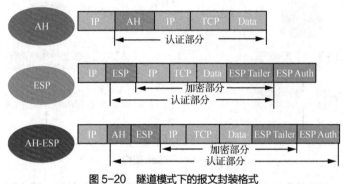

图 5-20　隧道模式下的报文封装格式

5.2.5　IPSec VPN 的配置流程

IPSec VPN 的配置流程如图 5-21 所示。

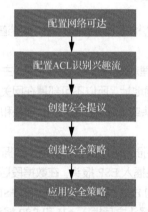

图 5-21　IPSec VPN 的配置流程

（1）配置网络可达：确保发送方与接收方之间的网络可达。检查报文发送方和接收方之间的网络层的连通性，确保双方只有建立 IPSec VPN 隧道后才能进行 IPSec 通信。

（2）配置 ACL 识别兴趣流：兴趣流是指网络中的特定流量，该流量表现出用户或系统的兴趣和需求，通常与某种特定的应用、协议或服务相关联。通过 ACL 来定义和区分不同的流量，有部分不需要满足完整性、机密性要求，所以要对流量进行过滤，选择出需要进行 IPSec 处理的兴趣流，通过配置 ACL 来定义和区分数据流。

（3）创建安全提议：为了能够正常传输数据流，安全隧道两端必须使用相同的认证算法、安全协议、加密算法和封装模式。如果要在两个安全网关之间建立 IPSec 安全隧道，则 IPSec 的隧道模式可以方便地隐藏通信过程中使用的实际源 IP 地址和目的 IP 地址。

（4）创建安全策略：每一个 IPSec 安全策略都使用唯一的名称和序号来标识。IPSec 策略中会应用

IPSec 提议中定义的封装模式、安全协议与认证算法。IPSec 安全策略可分成两类：手动建立 SA 的策略和 IKE 协商建立 SA 的策略（本书主要介绍手动建立 SA 的策略）。

（5）应用安全策略：在某一个端口上应用 IPSec 安全策略。对称加密算法建议使用 AES（128 位及以上密钥）。非对称加密算法建议使用 RSA（2048 位及以上密钥）。哈希算法建议使用 SHA-2（256 位及以上密钥）。密钥相关的哈希运算消息认证码（Hash-based Message Authentication Code，HMAC）算法建议使用 HMAC-SHA。

【任务实施】

实验 2　配置 IPSec VPN 技术

实验目的：
（1）掌握 IPSec 提议、策略、绑定方法。
（2）通过 ACL 定义兴趣流，经过 IPSec 过滤后的兴趣流将会对协商的各种参数进行处理并封装，之后通过 IPSec 隧道进行转发。

V5-5　配置 IPSec VPN 技术

配置公司总部到公司分部的 IPSec VPN，实现网络的互通。企业希望对分部与总部之间相互访问的流量进行安全保护。IPSec VPN 实验拓扑示意如图 5-22 所示，假设 R1 为分部网关，R3 为总部网关，R2 为公网网关。要求 R1 与 R3 之间通过公网建立通信，建立一条 IPSec 隧道来实施安全保护。安全策略实现过程如下。

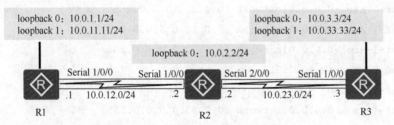

图 5-22　IPSec VPN 实验拓扑示意

（1）配置设备端口 IP 地址，代码如下。

```
[R1]interface Serial 1/0/0
[R1-Serial1/0/0]ip address 10.0.12.1 24
[R1-Serial1/0/0]quit
[R1]interface loopback 0
[R1-LoopBack0]ip address 10.0.1.1 24
[R1]interface loopback 1
[R1-LoopBack1]ip address 10.0.11.11 24

[R2]interface Serial 1/0/0
[R2-Serial1/0/0]ip address 10.0.12.2 24
[R2-Serial1/0/0]quit
[R2]interface Serial 2/0/0
```

```
[R2-Serial2/0/0]ip address 10.0.23.2 24
[R2]interface loopback 0
[R2-LoopBack0]ip address 10.0.2.2 24

[R3]interface Serial 2/0/0
[R3-Serial2/0/0]ip address 10.0.23.3 24
[R3-Serial2/0/0]quit
[R3]interface loopback 0
[R3-LoopBack0]ip address 10.0.3.3 24
[R3]interface loopback 1
[R3-LoopBack1]ip address 10.0.33.33 24
```

（2）配置 OSPF 路由，代码如下。

```
[R1]ospf router-id 10.0.1.1
[R1-ospf-1]area 0
[R1-ospf-1-area-0.0.0.0]network 10.0.12.0 0.0.0.255
[R1-ospf-1-area-0.0.0.0]network 10.0.1.0 0.0.0.255
[R1-ospf-1-area-0.0.0.0]network 10.0.11.0 0.0.0.255
[R2]ospf router-id 10.0.2.2
[R2-ospf-1]area 0
[R2-ospf-1-area-0.0.0.0]network 10.0.2.0 0.0.0.255
[R2-ospf-1-area-0.0.0.0]network 10.0.12.0 0.0.0.255
[R2-ospf-1-area-0.0.0.0]network 10.0.23.0 0.0.0.255
[R3]ospf router-id 10.0.3.3
[R3-ospf-1]area 0
[R3-ospf-1-area-0.0.0.0]network 10.0.23.0 0.0.0.255
[R3-ospf-1-area-0.0.0.0]network 10.0.3.0 0.0.0.255
[R3-ospf-1-area-0.0.0.0]network 10.0.33.0 0.0.0.255
```

（3）定义 ACL 的兴趣流，代码如下。

```
[R1-acl-adv-3001]rule 5 permit ip source 10.0.1.0 0.0.0.255 destination 10.0.3.0 0.0.0.255
[R3-acl-adv-3001]rule 5 permit ip source 10.0.3.0 0.0.0.255 destination 10.0.1.0 0.0.0.255
```

（4）创建安全提议，代码如下。

```
[R1-ipsec-proposal-tran1]esp authentication-algorithm sha1 [R1-ipsec-proposal-tran1]esp encryption-algorithm 3des
[R3-ipsec-proposal-tran1]esp authentication-algorithm sha1 [R3-ipsec-proposal-tran1]esp encryption-algorithm 3des
```

（5）创建安全策略，代码如下。

```
[R1]ipsec policy P1 10 manual
[R1-ipsec-policy-manual-P1-10]security acl 3001 [R1-ipsec-policy-manual-P1-10]
```

```
proposal tran1
    [R1-ipsec-policy-manual-P1-10]tunnel remote 10.0.23.3
    [R1-ipsec-policy-manual-P1-10]tunnel local 10.0.12.1
    [R1-ipsec-policy-manual-P1-10]sa spi outbound esp 54321 [R1-ipsec-policy-manual-P1-10]sa spi inbound esp 12345
    [R1-ipsec-policy-manual-P1-10]sa string-key outbound esp simple huawei [R1-ipsec-policy-manual-P1-10]sa string-key inbound esp simple huawei
    [R3]ipsec policy P1 10 manual
    [R3-ipsec-policy-manual-P1-10]security acl 3001 [R3-ipsec-policy-manual-P1-10]proposal tran1
    [R3-ipsec-policy-manual-P1-10]tunnel remote 10.0.12.1
    [R3-ipsec-policy-manual-P1-10]tunnel local 10.0.23.3
    [R3-ipsec-policy-manual-P1-10]sa spi outbound esp 12345 [R3-ipsec-policy-manual-P1-10]sa spi inbound esp 54321
    [R3-ipsec-policy-manual-P1-10]sa string-key outbound esp simple huawei [R3-ipsec-policy-manual-P1-10]sa string-key inbound esp simple huawei
```

（6）应用安全策略，代码如下。

```
    [R1]interface Serial 1/0/0
    [R1-Serial1/0/0]ipsec policy P1
    [R3]interface Serial 2/0/0
    [R3-Serial2/0/0]ipsec policy P1
```

（7）检测网络的连通性。

对非兴趣流不进行 IPSec 加密处理，代码如下。

```
    <R1>ping -a 10.0.11.11 10.0.33.33
    PING 10.0.33.33: 56 data bytes, press CTRL_C to break
    Reply from 10.0.33.33: bytes=56 Sequence=1 ttl=254 time=60 ms
    Reply from 10.0.33.33: bytes=56 Sequence=2 ttl=254 time=50 ms
    Reply from 10.0.33.33: bytes=56 Sequence=3 ttl=254 time=50 ms
    Reply from 10.0.33.33: bytes=56 Sequence=4 ttl=254 time=60 ms
    Reply from 10.0.33.33: bytes=56 Sequence=5 ttl=254 time=50 ms
    --------------------------------------------------------------------------------
    <R1>display ipsec statistics esp
    Inpacket count          0
    Inpacket auth count         0
    Inpacket decap count    0
    Outpacket count     0
    PktDuplicateDrop count      0
    PktSeqNoTooSmallDrop count0
    PktInSAMissDrop count       0
```

对兴趣流进行 IPSec 加密处理，代码如下。

```
<R1>ping -a 10.0.1.1 10.0.3.3
PING 10.0.3.3: 56 data bytes, press CTRL_C to break
Reply from 10.0.3.3: bytes=56 Sequence=1 ttl=255 time=80 ms
Reply from 10.0.3.3: bytes=56 Sequence=2 ttl=255 time=77 ms
Reply from 10.0.3.3: bytes=56 Sequence=3 ttl=255 time=77 ms
Reply from 10.0.3.3: bytes=56 Sequence=4 ttl=255 time=80 ms
Reply from 10.0.3.3: bytes=56 Sequence=5 ttl=255 time=77 ms
<R1>display ipsec statistics esp
-----------------------------------------------------------------
 Inpacket count             : 5
 Inpacket auth count        : 0
 Inpacket decap count       : 0
 Outpacket count            : 5
 PktDuplicateDrop count       0
 PktSeqNoTooSmallDrop count   0
 PktInSAMissDrop count        0
```

【任务拓展】

本实验分为两个阶段，第一阶段是建立管理连接，第二阶段是建立数据连接。请根据图 5-23 所示的拓扑示意，实现第一阶段和第二阶段的配置过程。

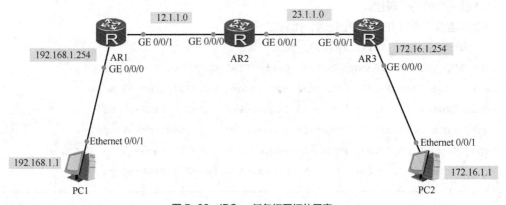

图 5-23　IPSec 任务拓展拓扑示意

第一阶段，在两个对等体设备之间建立一个安全的管理连接。没有实际的数据通过这个连接，这个管理连接是用来保护第二阶段的协商过程的。第一阶段需要协商的内容如下。

（1）双方使用什么加密算法（DES、3DES、AES）进行加密。

（2）摘要（完整性）认证的方式（MD5、SHA）。

（3）采用的密钥共享方式（预共享密钥、CA 数字签名、公钥认证）。

（4）使用的密钥强度 DH 组（密钥强度越大，表示加密强度越高）。

（5）管理连接生存时间（默认为一天，单位为 s）。

（6）协商模式（主模式或积极模式）。

第二阶段，当对等体设备之间有了安全的管理连接之后，它们就可以协商用于构建安全数据连接的安全参数了，这个协商过程是安全的、加密的。协商完成后，将在两个站点间形成安全的数据连接。第二阶段需要协商的内容如下。

（1）封装模式（是使用隧道模式还是使用传输模式）。

（2）封装技术（ESP、AH）。

（3）传输过程中数据的加密方式（DES、3DES、AES）。

（4）传输过程中数据的认证方式（MD5、SHA）。

（5）定义兴趣流（定义需要使用IPSec的流量）。

模块 6
网络服务技术

06

随着互联网的日益普及和飞速发展，接入互联网的局域网数量急剧增长，造成目前公有 IP 地址严重短缺，不可能保证每一台访问互联网的网络工作站均被分配到一个全球唯一的 IP 全局地址。为了节省公有 IP 地址，企业内部通常使用私有 IP 地址，通过在企业分支机构的边缘路由器上实现 NAT，使大量私有 IP 地址用户可以共享一个公有 IP 地址正常访问互联网。这样大型企业网络内部就会有大量的主机或设备需要获取私有 IP 地址等网络参数问题，如果采用手动配置，则工作量大且不好管理，如果有用户擅自修改网络参数，还有可能会造成 IP 地址冲突等问题。因此一般使用 DHCP 来分配 IP 地址等网络参数，可以减少管理员的工作量，避免用户手动配置网络参数时造成的地址冲突。本模块主要介绍网络服务技术，包括 NAT 服务器的配置、DHCP 的配置等。

【学习目标】

知识目标
（1）理解 NAT 技术的分类及工作流程。
（2）了解 DHCP 的应用场景。
（3）理解 DHCP 的报文类型及工作过程。
（4）理解 DHCP Relay 的原理。

技能目标
（1）熟悉常见 NAT 应用与处理常见的 NAT 问题。
（2）掌握 NAT 技术的静态配置与动态配置命令。
（3）重点掌握 NAPT 与 Easy IP 技术配置方法，在实际网络中灵活使用 NAT 技术。
（4）熟练掌握 DHCP 的具体应用和配置方法。
（5）掌握 DHCP 全局地址池的配置方法和端口地址池的配置方法。
（6）掌握在交换机端口上启用 DHCP 发现功能和 IP 地址分配功能的方法。
（7）掌握 DHCP Relay 的配置过程。

素质目标
（1）培养学生自主探究问题的精神。
（2）培养学生具备良好的网络工程意识，以及遵守国家法律和规章制度的意识。

任务 6.1　NAT 服务器的配置

【知识准备】

6.1.1　NAT 概述

NAT 的主要功能是通过转换报文中的 IP 地址，实现公网和私网的互访。NAT 通常部署在一个组织的网络出口位置，通过将内部网络 IP 地址替换为出口的 IP 地址提供公网可达性和上层协议的连接能力。

通常企业内部使用的是私有 IP 地址，然而，企业用户不仅需要访问私有网络（简称私网），还需要访问公有网络（简称公网）。图 6-1 所示为 NAT 服务示意。企业或家庭所使用的网络为私网，使用私有 IP 地址（如 10.0.0.1），私有网络内有两台主机，即 A(HostA) 和 B(HostB)；运营商维护的网络为公网，使用公有 IP 地址（如 198.76.28.11），NAT 一般部署在连接内网与外网的网关设备上。

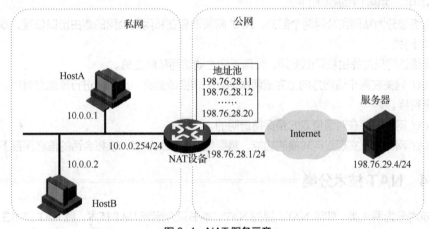

图 6-1　NAT 服务示意

6.1.2　私有 IP 地址和公有 IP 地址

1. 私有 IP 地址

私有 IP 地址是一段保留的 IP 地址，只使用在局域网中，无法在 Internet 上使用。IPv4 中私有 IP 地址范围：在 A 类地址中，10.0.0.0～10.255.255.255 是私有 IP 地址；在 B 类地址中，172.16.0.0～172.31.255.255 是私有 IP 地址；在 C 类地址中，192.168.0.0～192.168.255.255 是私有 IP 地址。这些 IP 地址可以在任何组织或企业内部使用，不能作为全球路由地址。对于有 Internet 访问需求而内部又使用私有 IP 地址的网络，就要在组织的出口位置部署 NAT 网关，在报文离开私网进入 Internet 时，将源 IP 地址替换为公有 IP 地址，通常是出口设备的端口 IP 地址。私有 IP 地址具有以下特点。

（1）任何组织都可以任意使用私有 IP 地址空间。

（2）私有 IP 地址在 Internet 上无法被识别。

（3）如果采用私有 IP 地址的网络需要访问 Internet，则必须在出口处部署 NAT 网关。

2. 公有 IP 地址

公有 IP 地址由互联网络信息中心（Internet Network Information Center，InterNIC）负责，分配给注册并向 InterNIC 提出申请的组织，通过它可直接访问 Internet，它是广域网范畴内的。组建一个企业级网络时，需要向 ISP 申请一个接入 Internet 的宽带，同时 ISP 会为其分配一个或多个 IP 地址，这些 IP 地址可以供企业内部网络联网，这些 ISP 分配的 IP 地址就是公有 IP 地址。IPv4 中公有 IP 地址的范围如下。

（1）A 类的公有 IP 地址：1.0.0.0～9.255.255.255，11.0.0.0～126.255.255.255。

（2）B 类的公有 IP 地址：128.0.0.0～172.15.255.255，172.32.0.0～191.255.255.255。

（3）C 类的公有 IP 地址：192.0.0.0～192.168.255.255，192.169.0.0～223.255.255.255。

6.1.3 NAT 工作流程

NAT 技术如何转换报文中的 IP 地址？从私网到公网：报文经过部署了 NAT 技术的路由器时，路由器将报文中的 SIP（源 IP 地址）由私有 IP 地址转换为公有 IP 地址。从公网到私网：报文经过部署了 NAT 技术的路由器时，路由器将报文中的 DIP（目的 IP 地址）由公有 IP 地址转换为私有 IP 地址。在整个 NAT 中，关键工作流程如下。

（1）网络被分为私网和公网两个部分，NAT 网关设置在私网到公网的路由出口位置，双向流量必须经过 NAT 网关。

（2）网络访问只能先由私网侧发起，公网无法主动访问私网主机。

（3）NAT 网关在两个访问方向上完成两次地址的转换或翻译，出方向进行源信息替换，入方向进行目的信息替换。

（4）NAT 网关的存在对通信双方是保持透明的。

（5）NAT 网关为了实现双向转换的功能，需要维护一张关联表，以将会话的信息保存下来。

6.1.4 NAT 技术分类

NAT 技术分为两大类，即源 NAT、目的 NAT。本书只介绍源 NAT 技术。源 NAT 技术主要有静态 NAT、基本 NAT、网络地址端口转换和 Easy IP 这 4 种类型。

1. 静态 NAT

所谓静态 NAT，是指公有 IP 地址和私有 IP 地址的对应关系是静态的，由管理员手动指定后就不再改变。根据图 6-1 可知，静态 NAT 实现的是一对一的地址转换，其私有 IP 地址和公有 IP 地址的对应关系如表 6-1 所示。显然，这种方式对节约公有 IP 地址没有太大意义，主要是为了实现一些特殊的组网需求，如用户希望隐藏内部主机的真实 IP 地址，或者实现两个 IP 地址重叠网络的通信。

表 6-1 静态 NAT 的私有 IP 地址和公有 IP 地址的对应关系

私有 IP 地址	公有 IP 地址
10.0.0.1	198.76.28.11
10.0.0.2	198.76.28.12
10.0.0.3	198.76.28.13
……	……

2. 基本 NAT

基本 NAT 仍然是一对一的地址转换,但是私有 IP 地址和公有 IP 地址的对应关系不固定。如表 6-2 所示,私有 IP 地址 10.0.0.1 可以对应公有 IP 地址池中剩余的任意 IP 地址,它是随机分配的。它无须静态配置路由器,使其将每个内部地址映射到一个外部地址,但必须有足够的公有 IP 地址,使连接 Internet 的主机都能够同时发送和接收分组。基本 NAT 是基于地址池来实现私有 IP 地址与公有 IP 地址转换的,基本 NAT 地址池中的地址用完以后,只能等待被占用的公有 IP 地址被释放后,其他主机才能使用它来访问公网。图 6-2 所示为基本 NAT 示意。

表 6-2 基本 NAT 的私有 IP 地址和公有 IP 地址的对应关系

私有 IP 地址	公有 IP 地址
10.0.0.1	198.76.28.11
10.0.0.2	198.76.28.12
10.0.0.3	198.76.28.13
……	……

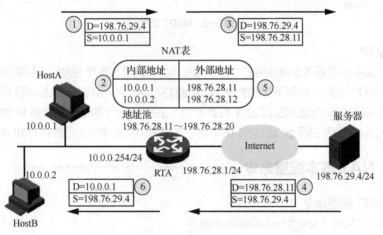

图 6-2 基本 NAT 示意

3. 网络地址端口转换

网络地址端口转换(Network Address Port Translation,NAPT)用于实现多对一的地址转换,将多个私有 IP 地址转换为同一个公有 IP 地址,允许多个私有 IP 地址映射到同一个公有 IP 地址的不同端口。在进行地址转换时,NAPT 不仅记录 IP 地址的转换关系,还记录端口号的对应关系,这样才能区分不同的私有 IP 地址。表 6-3 所示为 NAPT 方式下私有 IP 地址和公有 IP 地址的对应关系。

表 6-3 NAPT 方式下私有 IP 地址和公有 IP 地址的对应关系

私有 IP 地址	公有 IP 地址
10.0.0.1	198.76.28.11
10.0.0.2	
10.0.0.3	
……	

如表 6-3 所示,所有私有 IP 地址只对应一个公有 IP 地址 198.76.28.11,只需使用一个公有 IP 地址,就可将数千名用户连接到 Internet。其核心之处就在于利用端口号实现公有 IP 地址和私有 IP 地址的转

换。NAT 网关根据传输层信息或其他上层协议去区分不同的会话，并且可能要对上层协议的标识进行转换，如 TCP 或 UDP 端口号，这样 NAT 网关就可以将不同的内部连接访问映射到同一公有 IP 地址的不同传输层端口，通过这种方式实现公有 IP 地址的复用和解复用。图 6-3 所示为 NAPT 示意。

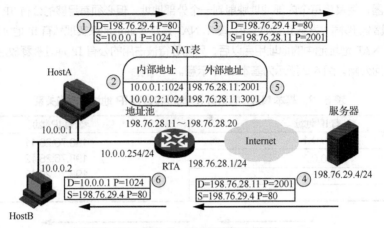

图 6-3　NAPT 示意

4. Easy IP

Easy IP 方式中 NAT 设备直接使用出口的 IP 地址作为转换后的源 IP 地址，不用预先配置地址池，工作原理与普通 NAPT 相同，也用于实现多对一的地址转换，将多个私有 IP 地址转换为同一个公有 IP 地址，转换后的公有 IP 地址为路由器出口的公有 IP 地址（见表 6-3），但转换后的公有 IP 地址（198.76.28.11 转换为 198.76.28.1）为路由器出口的公有 IP 地址，充分利用了路由器出口的公有 IP 地址的作用。

6.1.5　NAT 基本配置命令

1. 静态 NAT 配置命令

（1）启用静态 NAT（命令为 nat static enable）。

（2）绑定公有、私有 IP 地址（命令为 nat static global 公有 IP 地址 inside 私有 IP 地址）。其中，global 参数用于配置公有 IP 地址，inside 参数用于配置私有 IP 地址。建立公有 IP 地址与私有 IP 地址映射，如图 6-4 所示。

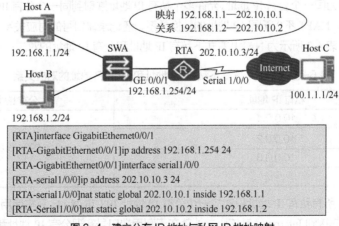

图 6-4　建立公有 IP 地址与私网 IP 地址映射

（3）静态 NAT 验证（命令为 display nat static）。如图 6-5 所示，Global IP/Port 表示公有 IP 地址和服务端口号，Inside IP/Port 表示私有 IP 地址和服务端口号。

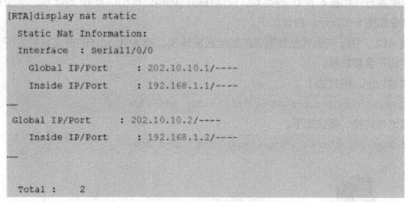

图 6-5　静态 NAT 验证

2. 基本 NAT 配置命令

基本 NAT 配置命令如图 6-6 所示，具体描述如下。

（1）配置 ACL，用于判断哪些数据包的地址应被转换，被 ACL 允许（permit）的报文将被进行 NAT 转换，被拒绝（deny）的报文将不会被转换。

（2）配置地址池（命令为 nat address-group），格式如下。

```
nat address-group group-number start-addr end-addr
```

（3）ACL 绑定地址池（命令为 nat outbound），用来将一个 ACL 和一个地址池关联起来，表示 ACL 中规定的地址可以使用地址池进行地址转换，no-pat 表示只转换数据报文的地址而不转换端口信息，格式如下。

```
nat outbound acl-number address-group group-number no-pat
```

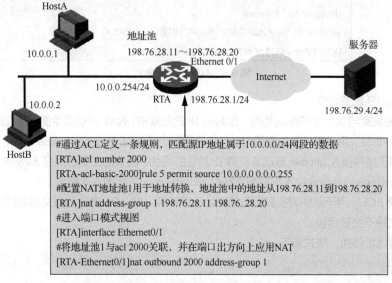

图 6-6　基本 NAT 配置命令

（4）查看基本 NAT 配置信息（命令为 display nat address-group）。

3. NAPT 配置命令

NAPT 的配置方式和基本 NAT 的类似，只是在最后调用公有和私有 IP 地址池时不加 no-pat 参数。NAPT 配置命令如图 6-7 所示，具体如下。

（1）配置 ACL，用于判断哪些数据包的地址应被转换，被 ACL 允许的报文将被进行 NAT 转换，被拒绝的报文将不会被转换。

（2）配置地址池，格式如下。

```
nat address-group group-number start-addr end-addr
```

（3）配置地址转换，格式如下。

```
nat outbound acl-number address-group group-number
```

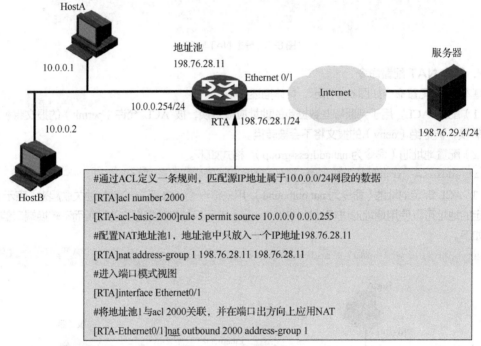

图 6-7　NAPT 配置命令

4. Easy IP 配置命令

Easy IP 的配置与 NAPT 的配置类似。在 Easy IP 的配置中，NAT 设备直接使用出口的 IP 地址作为转换后的源 IP 地址，用户不用预先配置地址池，只需要定义 ACL 和 nat outbound 命令，是 NAPT 的一种特例，适用于拨号接入 Internet 或动态获得 IP 地址的场合。两者的主要区别是 Easy IP 不需要配置地址池，所以 nat outbound 命令中不需要配置参数 address-group，如图 6-8 所示。

（1）配置 ACL，用于判断哪些数据包的地址应被转换，被 ACL 允许的报文将被进行 NAT 转换，被拒绝的报文将不会被转换。

（2）配置地址转换，格式如下。

```
nat outbound acl-number
```

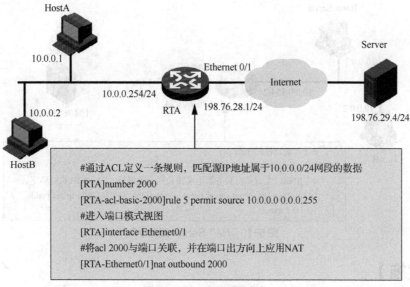

图 6-8 Easy IP 配置命令

5. NAT Server 配置命令

NAT Server 用于解决公网用户访问私网服务器的目的地址转换问题，如图 6-9 所示。当客户端想访问私网服务器或者私网服务器想访问客户端时，可以使用 NAT Server 实现，NAT Server 会将公有 IP 地址映射为私有 IP 地址。

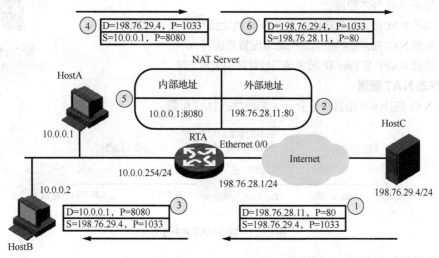

图 6-9 NAT Server 示意

NAT Server 配置命令如下。

```
nat server protocol pro-type global global-addr [global-port] inside host-addr
[host-port]
```

其中，global 后面是需要映射的公有 IP 地址或者公网端口；inside 后面是映射后的私有 IP 地址，因为 NAT Server 会产生 server-map 表，所以防火墙会先匹配 server-map 表，根据表项进行包装，但是 NAT Server 产生的 server-map 表不能绕过安全策略。NAT Server 配置命令如图 6-10 所示。

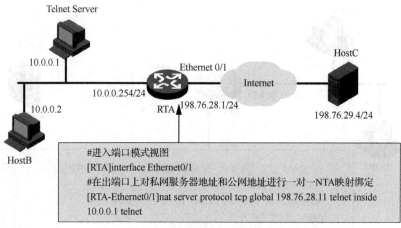

图 6-10　NAT Server 配置命令

【任务实施】

实验 1　配置 NAT Server

实验目的：
（1）理解 NAT 的应用场景。
（2）理解 NAT 技术原理。
（3）熟悉 NAT 技术如何转换报文中的 IP 地址及 NAT 技术的类型。
（4）掌握 NAT 技术的静态与动态地址配置方法。
（5）掌握 NAPT 与 Easy IP 技术端口地址转换配置方法。

V6-1　静态 NAT 配置

1. 静态 NAT 配置

静态 NAT 拓扑示意如图 6-11 所示，完成静态 NAT 配置。

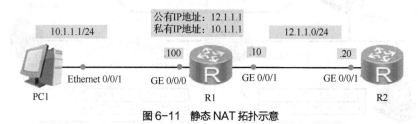

图 6-11　静态 NAT 拓扑示意

（1）配置端口 IP 地址，代码如下。

```
[R1]int g0/0/0
[R1-GigabitEthernet0/0/0]ip add 10.1.1.100 24
[R1-GigabitEthernet0/0/0]int g0/0/1
[R1-GigabitEthernet0/0/1]ip add 12.1.1.10 24


[R2]int g0/0/1
[R2-GigabitEthernet0/0/1]ip add 12.1.1.20 24
```

（2）在 PC1 端测试到 R2 的连通性，代码如下。

```
PC>ping 12.1.1.20


Ping 12.1.1.20: 32 data bytes, Press Ctrl_C to break
Request timeout!
Request timeout!
Request timeout!
Request timeout!
Request timeout!

--- 12.1.1.20 ping statistics ---
  5 packet(s) transmitted
  0 packet(s) received
  100.00% packet loss
```

结果显示是不通的，发送报文能够到达，而回送报文无法到达。

发送报文能到达，显示如下。

SIP：10.1.1.1 ------ 报文源地址（私有 IP 地址）
DIP：12.1.1.20 ------ 报文目的地址（公有 IP 地址）

回送报文无法到达，显示如下。

SIP：12.1.1.20 ------ 报文源地址（公有 IP 地址）
DIP：10.1.1.1 ------ 报文目的地址（私有 IP 地址）

（3）配置静态 NAT，启用静态 NAT，绑定公有、私有 IP 地址，代码如下。

```
[R1]int g0/0/1
[R1-GigabitEthernet0/0/1]nat static ?
  enable    Enable function
  global    Specify global information of NAT
  protocol  Specify protocol
[R1-GigabitEthernet0/0/1]nat static global ?
  X.X.X.X  Global IP address of NAT
[R1-GigabitEthernet0/0/1]nat static global 12.1.1.1 ?
  inside   Specify inside information of NAT
[R1-GigabitEthernet0/0/1]nat static global 12.1.1.1 inside ?
  IP_ADDR<X.X.X.X>  Inside IP address of NAT
[R1-GigabitEthernet0/0/1]nat static global 12.1.1.1 inside 10.1.1.1
[R1-GigabitEthernet0/0/1]
```

（4）在 PC1 端再次测试到 R2 的连通性。结果显示是通的，此时由于配置了静态 NAT，因此会把私有 IP 地址 10.1.1.1 转换为公有 IP 地址 12.1.1.1，显示如下。

```
PC>ping 12.1.1.20


Ping 12.1.1.20: 32 data bytes, Press Ctrl_C to break
```

```
From 12.1.1.20: bytes=32 seq=1 ttl=254 time=16 ms
From 12.1.1.20: bytes=32 seq=2 ttl=254 time=15 ms
From 12.1.1.20: bytes=32 seq=3 ttl=254 time=17 ms
From 12.1.1.20: bytes=32 seq=4 ttl=254 time=16 ms
```

（5）使用 display nat session all 命令查看静态 NAT 绑定信息，如图 6-12 所示。

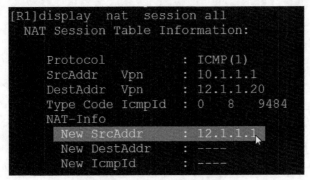

图 6-12 静态 NAT 绑定信息

V6-2 动态 NAT 配置

2. 动态 NAT 配置

动态 NAT 拓扑示意如图 6-13 所示，完成动态 NAT 配置。

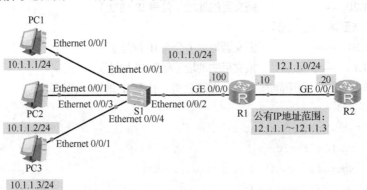

图 6-13 动态 NAT 拓扑示意

（1）设置端口 IP 地址，代码如下。

```
[R1]int g0/0/0
[R1-GigabitEthernet0/0/0]ip address 10.1.1.100 24
[R1-GigabitEthernet0/0/0]int g0/0/1
[R1-GigabitEthernet0/0/1]ip address 12.1.1.10 24

[R2]int g0/0/1
[R2-GigabitEthernet0/0/1]ip address 12.1.1.20 24
```

（2）建立地址池，代码如下。

```
[R1]nat address-group ?
  INTEGER<0-7>  Index of address-group
[R1]nat address-group 0 ?
```

```
  IP_ADDR<X.X.X.X>  Start address              //开始地址
[R1]nat address-group 0 12.1.1.1 ?
  IP_ADDR<X.X.X.X>  End address                //结束地址
[R1]nat address-group 0 12.1.1.1 12.1.1.3
#查看地址池
[R1]dis nat address-group 0
NAT Address-Group Information:
--------------------------------------
 Index   Start-address       End-address
--------------------------------------
 0       12.1.1.1            12.1.1.3
--------------------------------------
 Total : 1
```

（3）建立 ACL 规则，代码如下。

```
[R1]acl 2000
[R1-acl-basic-2000]rule permit source 10.1.1.1 0.0.0.0
[R1-acl-basic-2000]rule permit source 10.1.1.2 0.0.0.0
[R1-acl-basic-2000]rule permit source 10.1.1.3 0.0.0.0
[R1-acl-basic-2000]quit
```

或者使用如下代码。

```
[R1-acl-basic-2000]rule permit source 10.1.1.0 0.0.0.255
```

（4）ACL 绑定地址池，代码如下。

```
[R1]int GigabitEthernet 0/0/1
[R1-GigabitEthernet0/0/1]nat outbound ?
  INTEGER<2000-3999>  Apply basic or advanced ACL
[R1-GigabitEthernet0/0/1]nat outbound 2000 ?
  address-group  IP address-group of NAT
  interface      Specify the interface
  <cr>           Please press ENTER to execute command
[R1-GigabitEthernet0/0/1]nat outbound 2000 address-group ?
  INTEGER<0-7>  Index of address-group
[R1-GigabitEthernet0/0/1]nat outbound 2000 address-group 0 ?
  no-pat  Not use PAT
  <cr>    Please press ENTER to execute command
[R1-GigabitEthernet0/0/1]nat outbound 2000 address-group 0 no-pat
[R1-GigabitEthernet0/0/1]quit
```

（5）在 PC 端上测试到 R2 的连通性。在 PC1 上 ping 12.1.1.20，由于模拟器存在漏洞，地址池中只有 3 个公有 IP 地址，前 3 次发送报文占用了地址池的 3 个公有 IP 地址，后面发送的报文无法获取回送信息，如图 6-14 所示。

图 6-14 连通性测试

（6）查看 NAT 的转换表项。在 3 台 PC 上使用任意一台 PC ping 12.1.1.20，由于模拟器存在漏洞，地址池中的 3 个公有 IP 地址被 ping 的前 3 个报文占用（见图 6-14），PC3 发送到 12.1.1.20 的前 3 个报文把地址池中的 3 个公有 IP 地址抢占了。查看 NAT 的转换表项，如图 6-15 所示。

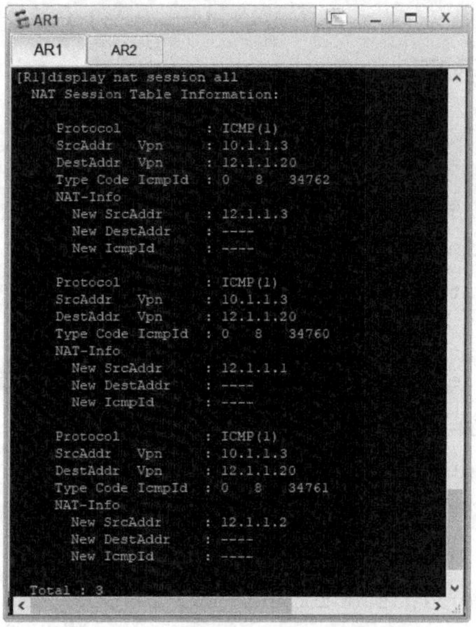

图 6-15 查看 NAT 的转换表项

3. NAPT 配置

NAPT 拓扑示意如图 6-16 所示。

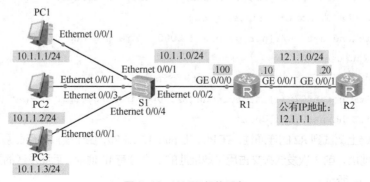

图 6-16 NAPT 拓扑示意

（1）建立地址池。在前面实验的基础上，使用 NAPT 实现多对一的地址转换，将多个私有 IP 地址转换为同一个公有 IP 地址。在进行地址转换时，NAPT 不仅记录了 IP 地址的转换关系，还记录了端口号的对应关系，这样才能区分不同的私有 IP 地址，具体代码如下。

```
[R1]nat address-group ?
  INTEGER<0-7>  Index of address-group
[R1]nat address-group 1 ?
  IP_ADDR<X.X.X.X>  Start address
[R1]nat address-group 1 12.1.1.1 ?
  IP_ADDR<X.X.X.X>  End address
[R1]nat address-group 1 12.1.1.1 12.1.1.1
#查看地址池
[R1]dis nat address-group 1

 NAT Address-Group Information:
 --------------------------------------
 Index   Start-address       End-address
 --------------------------------------
 1       12.1.1.1            12.1.1.1
 --------------------------------------
  Total : 1
```

（2）建立 ACL 规则，代码如下。

```
[R1]acl 2001
[R1-acl-basic-2001]rule permit source 10.1.1.1 0.0.0.0
[R1-acl-basic-2001]rule permit source 10.1.1.2 0.0.0.0
[R1-acl-basic-2001]rule permit source 10.1.1.3 0.0.0.0
```

或者使用如下代码。

```
[R1-acl-basic-2001]rule permit source 10.1.1.0 0.0.0.255
[R1-acl-basic-2001]dis this
[V200R003C00]
#
acl number 2001
 rule 5  permit source 10.1.1.1 0
 rule 10 permit source 10.1.1.2 0
 rule 15 permit source 10.1.1.3 0
 rule 20 permit source 10.1.1.0 0.0.0.255
#
return
[R1-acl-basic-2001]
```

（3）ACL 绑定地址池。NAPT 的配置方式和动态 NAT 的类似，只是在最后调用公有和私有 IP 地址池时不加 no-pat 参数即可。在之前实验的基础上，先删除 R1 的 GE 0/0/1 端口上绑定的地址池

acl 2000。重新在 GE 0/0/1 端口上绑定地址池 acl 2001，具体代码如下。

```
[R1-GigabitEthernet0/0/1]dis this
[V200R003C00]
#
interface GigabitEthernet0/0/1
 ip address 12.1.1.10 255.255.255.0
 nat outbound 2000 address-group 0 no-pat
#
return
[R1-GigabitEthernet0/0/1]undo nat outbound 2000 address-group 0 no-pat
[R1-GigabitEthernet0/0/1]
[R1-GigabitEthernet0/0/1]nat outbound 2001 address-group 1 ?
  no-pat  Not use PAT
  <cr>    Please press ENTER to execute command
[R1-GigabitEthernet0/0/1]nat outbound 2001 address-group 1
[R1-GigabitEthernet0/0/1]
```

（4）测试并查看 NAT。分别在 3 台 PC 上使用 ping 12.1.1.20 命令，查看 NAT 的转换表项得知，每个报文的私有 IP 地址都转换为一个新公有 IP 地址（New ScrAddr：12.1.1.1），具体代码如下。

```
[R1]dis nat session all
  NAT Session Table Information:
     Protocol         : ICMP(1)
  SrcAddr  Vpn      : 10.1.1.3
     DestAddr  Vpn  : 12.1.1.20
     Type Code IcmpId : 0  8  36310
     NAT-Info
 New SrcAddr      : 12.1.1.1
     New DestAddr   : ----
     New IcmpId     : 10249


     Protocol         : ICMP(1)
 SrcAddr  Vpn       : 10.1.1.1
     DestAddr  Vpn  : 12.1.1.20
     Type Code IcmpId : 0  8  36310
     NAT-Info
 New SrcAddr      : 12.1.1.1
     New DestAddr   : ----
     New IcmpId     : 10248


     Protocol         : ICMP(1)
 SrcAddr  Vpn       : 10.1.1.2
```

```
    DestAddr    Vpn       : 12.1.1.20
    Type Code IcmpId      : 0   8    36316
    NAT-Info
New SrcAddr       : 12.1.1.1
    New DestAddr      : ----
    New IcmpId        : 10255
```

4. Easy IP 配置

Easy IP 的工作方式和 NAPT 的类似，转换后的公有 IP 地址为路由器出口的公有 IP 地址。Easy IP 拓扑示意如图 6-17 所示，完成 Easy IP 的配置。Easy IP 不需要建立地址池，直接使用 R1 的 GE 0/0/1 端口的 IP 地址 12.1.1.10 作为内网的公有 IP 地址。只需要创建一个普通的 ACL 规则，并在 R1 的 GE 0/0/1 端口上绑定该 ACL 即可。

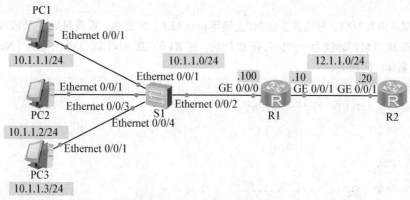

图 6-17　Easy IP 拓扑示意

（1）配置 Easy IP，代码如下。

```
[R1]acl 2002
[R1-acl-basic-2002]rule permit source 10.1.1.0 0.0.0.255
[R1-acl-basic-2002]rule permit source 10.1.1.1 0.0.0.0
[R1-acl-basic-2002]rule permit source 10.1.1.2 0.0.0.0
[R1-acl-basic-2002]rule permit source 10.1.1.3 0.0.0.0
[R1-acl-basic-2002]dis this
[V200R003C00]
#
acl number 2002
 rule 5 permit source 10.1.1.0 0.0.0.255
 rule 10 permit source 10.1.1.1 0
 rule 15 permit source 10.1.1.2 0
 rule 20 permit source 10.1.1.3 0
#
return
[R1-acl-basic-2002]quit
```

```
[R1]int g0/0/1
[R1-GigabitEthernet0/0/1]nat ou
[R1-GigabitEthernet0/0/1]nat outbound 2002
[R1-GigabitEthernet0/0/1]undo nat outbound 2001 address-group 1
[R1-GigabitEthernet0/0/1]dis this
[V200R003C00]
#
interface GigabitEthernet0/0/1
ip address 12.1.1.10 255.255.255.0
nat outbound 2002
#
Return
```

（2）测试并查看 NAT。分别在 3 台 PC 上使用 ping 12.1.1.20 命令，查看 NAT 的转换表项得知，每个报文的私有 IP 地址都转换为一个新公有 IP 地址，即 R1 的 GE 0/0/1 端口的 IP 地址（New ScrAddr: 12.1.1.10），具体代码如下。

```
[R1]dis nat session all
 NAT Session Table Information:

   Protocol        : ICMP(1)
 SrcAddr   Vpn    : 10.1.1.3
    DestAddr  Vpn    : 12.1.1.20
    Type Code IcmpId : 0  8   37745
    NAT-Info
 New SrcAddr     : 12.1.1.10
    New DestAddr    : ----
    New IcmpId      : 10255


   Protocol        : ICMP(1)
 SrcAddr   Vpn    : 10.1.1.2
    DestAddr  Vpn    : 12.1.1.20
    Type Code IcmpId : 0  8   37742
    NAT-Info
 New SrcAddr     : 12.1.1.10
    New DestAddr    : ----
    New IcmpId      : 10254


   Protocol        : ICMP(1)
 SrcAddr   Vpn    : 10.1.1.1
    DestAddr  Vpn    : 12.1.1.20
    Type Code IcmpId : 0  8   37739
```

```
        NAT-Info
New SrcAddr      : 12.1.1.10
    New DestAddr    : ----
    New IcmpId      : 10251
```

5. NAT Server 配置

配置 NAT Server,可以使外网用户访问内网服务器。NAT Server 拓扑示意如图 6-18 所示,完成 NAT Server 的配置。

V6-3 NAT server 配置

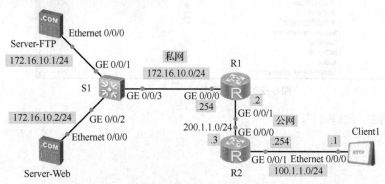

图 6-18 NAT Server 拓扑示意

(1)根据拓扑完成编址,代码如下。

```
[R1]int g0/0/0
[R1-GigabitEthernet0/0/0]ip add 172.16.10.254 24
[R1-GigabitEthernet0/0/0]int g0/0/1
[R1-GigabitEthernet0/0/1]ip add 200.1.1.2 24


[R2]int g0/0/0
[R2-GigabitEthernet0/0/0]ip add 200.1.1.3 24
[R2-GigabitEthernet0/0/0]int g0/0/1
[R2-GigabitEthernet0/0/1]ip add 100.1.1.254 24
```

(2)在私网中设置 Web 与 FTP 服务,如图 6-19 和图 6-20 所示。

图 6-19 Web 服务

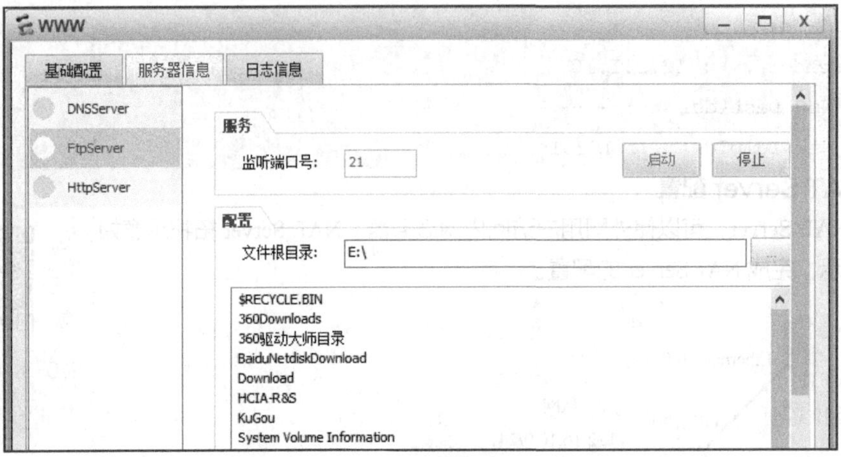

图 6-20 FTP 服务

（3）在公网内实现路由，代码如下。

```
[R1]rip
[R1-rip-1]network 200.1.1.0


[R2]rip
[R2-rip-1]network 200.1.1.0
[R2-rip-1]network 100.0.0.0
[R2-rip-1]dis this
[V200R003C00]
#
rip 1
 network 200.1.1.0
 network 100.0.0.0
#
return
[R2-rip-1]qu
```

（4）在 R1 上配置 NAT Server，代码如下。

```
[R1]int g0/0/1
[R1-GigabitEthernet0/0/1]nat server protocol tcp global ?
  X.X.X.X            Global IP address of NAT
  current-interface  Address of current interface
  interface          Specify the interface
[R1-GigabitEthernet0/0/1]nat server protocol tcp global current-interface ftp inside 172.16.10.1 21
Warning:The port 21 is well-known port. If you continue it may cause function failure.
Are you sure to continue?[Y/N]:y
```

```
[R1-GigabitEthernet0/0/1]dis this
[V200R003C00]
#
interface GigabitEthernet0/0/1
 ip address 200.1.1.2 255.255.255.0
 nat server protocol tcp global current-interface ftp inside 172.16.10.1 ftp
#
return
[R1-GigabitEthernet0/0/1]nat server protocol tcp global current-interface www in
side 172.16.10.2 80
Warning:The port 80 is well-known port. If you continue it may cause function fa
ilure.
Are you sure to continue?[Y/N]:y
[R1-GigabitEthernet0/0/1]dis this
[V200R003C00]
#
interface GigabitEthernet0/0/1
 ip address 200.1.1.2 255.255.255.0
 nat server protocol tcp global current-interface ftp inside 172.16.10.1 ftp
 nat server protocol tcp global current-interface www inside 172.16.10.2 www
#
return
[R1-GigabitEthernet0/0/1]qu
```

（5）查看 NAT 信息，代码如下。

```
[R1]dis nat server

  Nat Server Information:
  Interface  : GigabitEthernet0/0/1
    Global IP/Port     : current-interface/21(ftp) (Real IP : 200.1.1.2)
    Inside IP/Port     : 172.16.10.1/21(ftp)
    Protocol : 6(tcp)
    VPN instance-name  : ----
    Acl number         : ----
    Description :  ----

    Global IP/Port     : current-interface/80(www) (Real IP : 200.1.1.2)
    Inside IP/Port     : 172.16.10.2/80(www)
    Protocol : 6(tcp)
    VPN instance-name  : ----
    Acl number         : ----
```

```
    Description : ----

    Total :    2
```

（6）公网站点访问私网服务器，如图 6-21 所示。

图 6-21　Web 服务访问

【任务拓展】

根据图 6-22 所示的拓扑示意，完成私网设备通过 NAT 后可以访问公网服务器，公网客户端通过 NAT Server 可以访问私网服务器的任务。

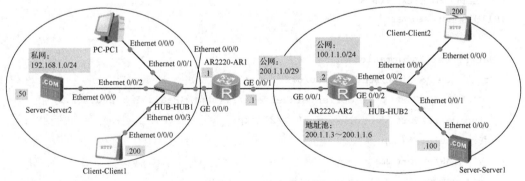

图 6-22　NAT 拓扑示意

任务 6.2　DHCP 的配置

【知识准备】

动态主机配置协议（Dynamic Host Configuration Protocol，DHCP）是由 IETF 制定的，是一个简化

主机 IP 地址分配管理的 TCP/IP，使用 UDP 工作，主要给内部网络或网络服务提供商自动分配 IP 地址。DHCP 有 3 个端口，其中 UDP:67 和 UDP:68 为正常的 DHCP 端口，分别作为 DHCP 服务器和 DHCP 客户端的服务端口。

6.2.1 DHCP 概述

DHCP 可以为客户端主机自动分配 TCP/IP 参数信息。网络中的主机作为 DHCP 客户端时，可以从网络中的 DHCP 服务器下载网络的配置信息，这些信息包括 IP 地址、子网掩码、网关、DNS 服务器和代理服务器的 IP 地址。通过 DHCP 服务器，不再需要手动设定网络配置信息，从而给网络集中管理不同的系统带来了方便。网络管理员可以通过配置 DHCP 服务器来实现对网络中不同系统的网络配置。

假如你是公司的网络管理员，公司网络需要配置 DHCP 业务，将网关路由器配置为 DHCP 服务器，并配置全局地址池和端口地址池，为接入层设备分配 IP 地址，那么你可能第一时间会使用 DHCP 配置去实现。DHCP 是一种客户端/服务器（Client/Server，C/S）架构的协议，通过 DHCP 可以使 DHCP 的客户端从服务器获取到 IP 地址等网络信息，如图 6-23 所示。

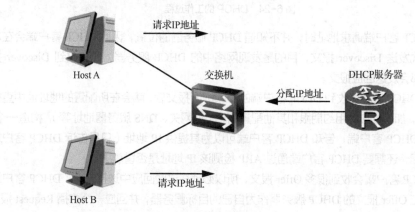

图 6-23 DHCP 服务

6.2.2 DHCP 的报文类型和工作过程

1. DHCP 的报文类型

DHCP 的报文主要有以下几种类型。

（1）Discover 报文：DHCP 客户端用来寻找 DHCP 服务器。

（2）Offer 报文：DHCP 服务器用来响应 Discover 报文，此报文携带了各种配置信息。

（3）Request 报文：DHCP 客户端请求配置确认，或者续借租期。

（4）ACK 报文：DHCP 服务器对 Request 报文的确认响应。

（5）Nak 报文：DHCP 服务器对 Request 报文的拒绝响应。

（6）Release 报文：当 DHCP 客户端要释放地址时，此报文用来通知 DHCP 服务器。

（7）Decline 报文：分配的 IP 地址不可使用时，通知 DHCP 服务器。

（8）Inform 报文：如果 DHCP 客户端需要从 DHCP 服务器获取更为详细的配置信息，则发送 Inform 报文向 DHCP 服务器进行请求，DHCP 服务器收到该报文后，将根据租约进行查找，找到相应的配置信息后，发送 ACK 报文回应 DHCP 客户端（极少用到）。

2. DHCP 的工作过程

DHCP 的工作过程如图 6-24 所示。

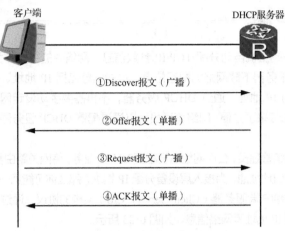

图 6-24　DHCP 的工作过程

首先，DHCP 客户端请求地址时，并不知道 DHCP 服务器的位置，因此 DHCP 客户端会在本地网络内以广播方式发送 Discover 报文，目的是发现网络中的 DHCP 服务器，所有收到 Discover 报文的 DHCP 服务器都会发送回应报文。

网络中的 DHCP 服务器收到 DHCP 客户端的 Discover 报文后，就会在所配置的地址池中查找一个合适的 IP 地址，加上相应的租约期限和其他配置信息（如网关、DNS 服务器地址等），构造一个 Offer 报文，发送给 DHCP 客户端，告知 DHCP 客户端可以为其提供 IP 地址（只是告诉 DHCP 客户端可以提供，是预分配，还需要 DHCP 客户端通过 ARP 检测该 IP 地址是否重复）。

其次，DHCP 客户端会收到很多 Offer 报文，所以必须在这些回应中选择一个。DHCP 客户端通常选择第一个回应 Offer 报文的 DHCP 服务器作为自己的目标服务器，并回应一个广播 Request 报文，通告选择的 DHCP 服务器。

DHCP 服务器收到 Request 报文后，根据 Request 报文中携带的用户 MAC 地址来查找有没有相应的续约记录，如果有则发送 ACK 报文作为回应，通知用户可以使用分配的 IP 地址。

如果 DHCP 服务器收到 Request 报文后，没有发现相应的租约记录或者由于某些原因无法正常分配 IP 地址，则发送 ACK 报文作为回应，通知用户无法分配合适的 IP 地址。

DHCP 客户端收到 DHCP 服务器回应的 ACK 报文后，若通过地址冲突检测发现 DHCP 服务器分配的地址冲突或者由于其他原因导致不能使用，则发送 Decline 报文，通知 DHCP 服务器所分配的 IP 地址不可用。

最后，DHCP 客户端成功获取 IP 地址后，在地址使用租期过去 1/2 时，会向 DHCP 服务器发送单播 Request 报文续延租期，如图 6-25（a）所示。如果没有收到 ACK 报文，则在租期过去 3/4 时，DHCP 客户端会发送广播 Request 报文续延租期，如图 6-25（b）所示。当用户不再需要使用分配的 IP 地址时，就会向 DHCP 服务器发送 Release 报文，告知 DHCP 服务器用户不再需要分配 IP 地址，DHCP 服务器会释放被绑定的租约。若 IP 地址租约到期前都没有收到 DHCP 服务器响应，则 DHCP 客户端停止使用此 IP 地址，DHCP 客户端也可主动向 DHCP 服务器发送 Release 报文，请求释放 IP 地址，如图 6-25（c）所示。

(a)向 DHCP 服务器发送单播 Request 报文续延租期

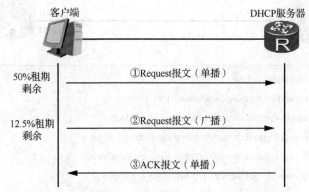

(b)向 DHCP 服务器发送广播 Request 报文续延租期

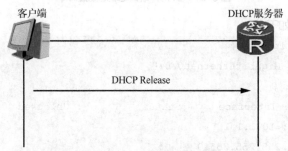

(c)DHCP 客户端主动向 DHCP 服务器发送 Release 报文,请求释放 IP 地址

图 6-25　不同 IP 地址租约期限的服务器反应

6.2.3　DHCP 基本配置命令

ARG3 系列路由器支持两种地址池,即全局地址池和端口地址池,如图 6-26 所示。

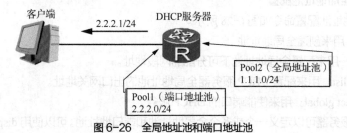

图 6-26　全局地址池和端口地址池

1. DHCP 端口地址池配置

DHCP 端口地址池配置命令如图 6-27 所示。

（1）dhcp enable：用来启用 DHCP 服务。
（2）dhcp select interface：用来关联端口和端口地址池，为连接到端口的主机提供配置信息。
（3）dhcp server dns-list：用来指定端口地址池下的 DNS 服务器地址。
（4）dhcp server excluded-ip-address：用来配置端口地址池中不参与自动分配的 IP 地址范围。
（5）dhcp server lease：用来配置租期。

```
[Huawei]dhcp enable
[Huawei]interface GigabitEthernet0/0/0
[Huawei-GigabitEthernet0/0/0]dhcp select interface
[Huawei-GigabitEthernet0/0/0]dhcp server dns-list 10.1.1.2
[Huawei-GigabitEthernet0/0/0]dhcp server excluded-ip-address 10.1.1.2
[Huawei-GigabitEthernet0/0/0]dhcp server lease day 3
```

图 6-27　DHCP 端口地址池配置命令

查看地址池的代码如下。

```
[Huawei]display ip pool
 Pool-name        : GigabitEthernet0/0/0
 Pool-No          : 0
 Position         : Interface       Status          : Unlocked
 Gateway-0        : 10.1.1.1
 Mask             : 255.255.255.0
 VPN instance     : --
 IP address Statistic
   Total          :253
   Used           :1         Idle      :252
   Expired        :0         Conflict  :0         Disable  :1
```

2. DHCP 全局地址池配置

DHCP 全局地址池配置命令如图 6-28 所示。

（1）ip pool：用来创建全局地址池。
（2）network：用来配置全局地址池下可分配的网段地址。
（3）gateway-list：用来配置 DHCP 服务器全局地址池的出口网关地址。
（4）dhcp select global：用来使能端口的 DHCP 服务。

每台 DHCP 服务器可以定义一个或多个全局地址池和端口地址池。可以使用 display ip pool 命令来查看地址池。

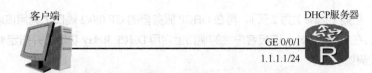

```
            [Huawei]dhcp enable
            [Huawei]ip pool pool2
            Info：It's successful to create an IP address pool.
            [Huawei-ip-pool-pool2]network 1.1.1.0 masd 24
            [Huawei-ip-pool-pool2]gateway-list 1.1.1.1
            [Huawei-ip-pool-pool2]lease day 10
            [Huawei-ip-pool-pool2]quit
            [Huawei]interface GigabitEthernet0/0/1
            [Huawei-GigabitEthernet0/0/1]dhcp select global
```

图 6-28　DHCP 全局地址池配置命令

查看地址池的代码如下。

```
[Huawei]display ip pool
------------------------------------------------------------
 Pool-name         : pool2
 Pool-No           : 0
 Position          : Local          Status         : Unlocked
 Gateway-0         : 1.1.1.1
 Mask              : 255.255.255.0
 VPN instance      : --
 IP address Statistic
   Total    :253
   Used     :1         Idle       :252
   Expired  :0         Conflict   :0        Disable   :0
```

6.2.4　DHCP Relay

1. 应用场景

当网络内有多个网段需要使用 DHCP 服务时，因为在 IP 地址动态获取过程中，客户端采取广播方式发送请求报文，而广播报文不能跨网段传送，所以 DHCP 只适用于客户端和服务器处于同一网段内的情况，当多个网段需要进行动态 IP 地址分配时，就需要在所有网段上设置一台 DHCP 服务器，这显然不易于管理和维护。这时可以通过 DHCP Relay（DHCP 中继）技术实现集中管理，方便 IP 地址分配。

DHCP Relay 可以使 DHCP 客户端通过它与其他网段的 DHCP 服务器通信，最终获取 IP 地址。路由器或者三层交换机都可以充当 DHCP Relay 设备。

2. 配置 DHCP Relay

DHCP Relay 配置如图 6-29 所示，先配置 DHCP 服务器，在 DHCP 服务器上配置客户端对应的 DHCP 地址池[网络号（network）为 10.1.1.0/24，网关（gateway）为 10.1.1.1，域名服务器（dns-server）

地址为 10.1.1.1，租期（lease）为 2 天]，再在 DHCP 服务器的 GE 0/0/2 端口上调用该地址池，最后配置 DHCP Relay，在 DHCP Relay 连接客户端的端口上启用 DHCP Relay 功能，并指定 DHCP 服务器的 IP 地址。具体代码如下。

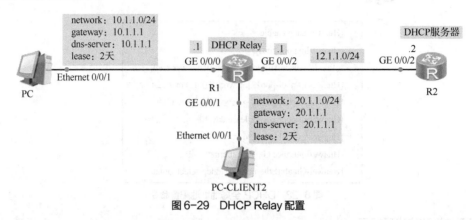

图 6-29　DHCP Relay 配置

```
[R1]dhcp enable
[R1]int g0/0/0
[R1-GigabitEthernet0/0/0]dhcp select relay
[R1-GigabitEthernet0/0/0]dhcp relay server-ip 12.1.1.2
[R1-GigabitEthernet0/0/0]quit
[R1]int g0/0/1
[R1-GigabitEthernet0/0/1]dhcp  select relay
[R1-GigabitEthernet0/0/1]dhcp  relay server-ip 12.1.1.2
```

3. DHCP Relay 的工作机制

配置 DHCP Relay 后，客户端通过广播将 Discover 报文发送给 DHCP Relay。DHCP Relay 在收到客户端的 Discover 报文后，Discover 报文重新被封装。源 IP 地址（SIP）被封装为接收报文的端口 IP 地址，目的 IP 地址（DIP）被封装为 DHCP 服务器的端口 IP 地址。将 Discover 报文通过单播方式发送给 DHCP 服务器。DHCP 服务器在收到 DHCP Relay 发送来的 Discover 报文后，通过单播方式向 DHCP Relay 回复 Offer 报文。DHCP Relay 收到 DHCP 服务器发送来的 Offer 报文后，将单播 Offer 报文转发给客户端。

【任务实施】

实验 2　配置 DHCP

实验目的：
（1）掌握 DHCP 的基本原理。
（2）掌握 DHCP 的应用场景。
（3）熟悉 DHCP 的基本配置过程。

V6-4　配置 DHCP

1. 实验拓扑

DHCP 服务配置拓扑示意如图 6-30 所示，完成 R1 上 DHCP 服务的配置，要求 3 台 PC 能自动获取 10.1.1.0/24 网段内的 IP 地址和网关为 10.1.1.1/24 的 IP 地址。

2. 实验过程

（1）配置 R1 端口的 IP 地址，代码如下。

```
[R1]int g0/0/0
[R1-GigabitEthernet0/0/1]ip add 10.1.1.1 24
```

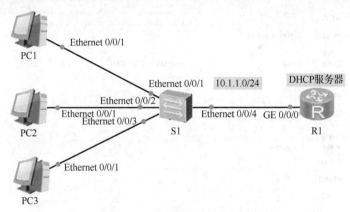

图 6-30　DHCP 服务配置拓扑示意

（2）启用 DHCP 服务，代码如下。

```
[R1]dhcp enable
```

（3）建立地址池，代码如下。

```
[R1]ip pool dhcp1                                   //定义地址池名称
[R1-ip-pool-dhcp1]network 10.1.1.0 mask 255.255.255.0    //创建地址池
[R1-ip-pool-dhcp1]gateway-list 10.1.1.1             //网关
[R1-ip-pool-dhcp1]dns-list 10.1.1.1                 //域名
[R1-ip-pool-dhcp1]lease day 2                       //租期
[R1-ip-pool-dhcp1]quit

[R1-ip-pool-dhcp1]display this
[V200R003C00]
#
ip pool dhcp1
 gateway-list 10.1.1.1
 network 10.1.1.0 mask 255.255.255.0
 lease day 2 hour 0 minute 0
 dns-list 10.1.1.1
#
return
[R1-ip-pool-dhcp1]
```

（4）在端口上启用地址池，代码如下。

```
[R1]int g0/0/0
```

```
[R1-GigabitEthernet0/0/2]dhcp select global
[R1-GigabitEthernet0/0/2]quit
```

（5）在 3 台 PC 上设置其自动获取 IP 地址，如图 6-31 所示。

图 6-31　设置 PC 自动获取 IP 地址

查询某 PC 获取的 IP 地址信息，结果如图 6-32 所示。

图 6-32　IP 地址信息查询结果

（6）验证 DHCP 服务信息，代码如下。

```
[R1]display ip pool name dhcp1
  Pool-name         : dhcp1
  Pool-No           : 0
  Lease             : 2 Days 0 Hours 0 Minutes
  Domain-name       : -
  DNS-server0       : 10.1.1.1
  NBNS-server0      : -
  Netbios-type      : -
  Position          : Local           Status            : Unlocked
```

```
 Gateway-0          : 10.1.1.1
 Mask               : 255.255.255.0
 VPN instance       : --
--------------------------------------------------------------------
 Start              End            Total   Used   Idle(Expired)   Conflict   Disable
--------------------------------------------------------------------
 10.1.1.1           10.1.1.254     253     3      250(0)          0          0
--------------------------------------------------------------------
```

实验 3　配置 DHCP Relay

实验目的：

（1）掌握通过 DHCP Relay 实现跨网络的 DHCP 服务的方法。

（2）掌握 DHCP Relay 的工作原理和工作方法。

（3）掌握 DHCP Relay 的配置过程。

V6-5　配置 DHCP Relay

1. 实验拓扑

完成 DHCP Relay 配置，要求 PC1、PC2 能分别自动获取 192.168.10.0/24 网段和 192.168.20.0/24 网段内的 IP 地址，拓扑示意如图 6-33 所示。

图 6-33　DHCP Relay 配置拓扑示意

2. 实验过程

（1）配置交换机，代码如下。

```
//创建 VLAN，并将端口加入 VLAN
[SW1]vlan batch 10 20
[SW1]int g0/0/1
[SW1-GigabitEthernet0/0/1]port link-type access
[SW1-GigabitEthernet0/0/1]port default vlan 10
[SW1-GigabitEthernet0/0/1]int g0/0/2
[SW1-GigabitEthernet0/0/2]port link-type access
[SW1-GigabitEthernet0/0/2]port default vlan 20
[SW1-GigabitEthernet0/0/2]int g0/0/3
```

```
[SW1-GigabitEthernet0/0/3]port link-type access
[SW1-GigabitEthernet0/0/3]port default vlan 20
//设置VLANIF地址
[SW1]int Vlanif 1
[SW1-Vlanif1]ip add 10.10.10.1 24
[SW1-Vlanif1]int Vlanif 10
[SW1-Vlanif10]ip add 192.168.10.1 24
[SW1-Vlanif10]int Vlanif 20
[SW1-Vlanif20]ip add 192.168.20.1 24
```

（2）设置路由器端口的IP地址，代码如下。

```
[R1]int g0/0/0
[R1-GigabitEthernet0/0/0]ip add 10.10.10.2 24
[R1-GigabitEthernet0/0/0]int g0/0/1
[R1-GigabitEthernet0/0/1]ip add 20.20.20.1 24

[R2]int g0/0/1
[R2-GigabitEthernet0/0/1]ip add 20.20.20.2 24
```

（3）配置动态路由，代码如下。

```
[SW1]rip
[SW1-rip-1]network 10.0.0.0
[SW1-rip-1]network 192.168.10.0
[SW1-rip-1]network 192.168.20.0

[R1]rip
[R1-rip-1]network 10.0.0.0
[R1-rip-1]network 20.0.0.0

[R2]rip
[R2-rip-1]network 20.0.0.0
```

（4）配置DHCP服务器，代码如下。

```
//创建两个地址池，这里只给出具体命令
ip pool p10
 gateway-list 192.168.10.1
 network 192.168.10.0 mask 255.255.255.0
 lease day 2 hour 0 minute 0
 dns-list 1.1.1.1 2.2.2.2
ip pool p20
 gateway-list 192.168.20.1
 network 192.168.20.0 mask 255.255.255.0
 excluded-ip-address 192.168.20.200 192.168.20.254
```

```
 lease day 0 hour 2 minute 0
 dns-list 1.1.1.1 2.2.2.2
[R2]dhcp enable
[R2]ip pool p10
[R2-ip-pool-p10]gateway-list 192.168.10.1
[R2-ip-pool-p10]network 192.168.10.0 mas
[R2-ip-pool-p10]network 192.168.10.0 mask 255.255.255.0
[R2-ip-pool-p10]lease day 2 hour 0 minute 0
[R2-ip-pool-p10]dns-list 1.1.1.1 2.2.2.2
[R2-ip-pool-p10]qu
[R2]ip pool p20
[R2-ip-pool-p20]gateway-list 192.168.20.1
[R2-ip-pool-p20]network 192.168.20.0 mask 255.255.255.0
[R2-ip-pool-p20]excluded-ip-address 192.168.20.200 192.168.20.254
[R2-ip-pool-p20]lease day 0 hour 2 minute 0
[R2-ip-pool-p20]dns-list 1.1.1.1 2.2.2.2
[R2-ip-pool-p20]
//在GE 0/0/0端口上启用全局DHCP地址池
[R2]int g0/0/1
[R2-GigabitEthernet0/0/1]dhcp select global
```

（5）配置DHCP Relay，代码如下。

```
[SW1]dhcp enable
[SW1]int Vlanif 10
[SW1-Vlanif10]dhcp select relay
[SW1-Vlanif10]dhcp relay server-ip 20.20.20.2

[SW1]int Vlanif 20
[SW1-Vlanif20]dhcp select relay
[SW1-Vlanif20]dhcp relay server-ip 20.20.20.2
```

（6）在PC端进行验证。在PC1上设置通过DHCP服务器获取IP地址，如图6-34所示。

图6-34 在PC1上设置通过DHCP服务器获取IP地址

PC1获取的IP地址如图6-35所示。

图 6-35　PC1 获取的 IP 地址

在 PC2 上设置通过 DHCP 服务器获取 IP 地址，获取的 IP 地址如图 6-36 所示。

图 6-36　PC2 获取的 IP 地址

【任务拓展】

根据图 6-37 所示的拓扑示意，完成客户端 PC1、PC2、PC4、PC5 动态获取 IP 地址的功能，在 AR4 上配置 DHCP 服务器，创建 4 个 IP 地址池（地址池名称分别是 V10、V20、V30、V40），在 LSW1 和 AR3 上配置 DHCP Relay。

V6-6　任务 6.2 拓展任务提示

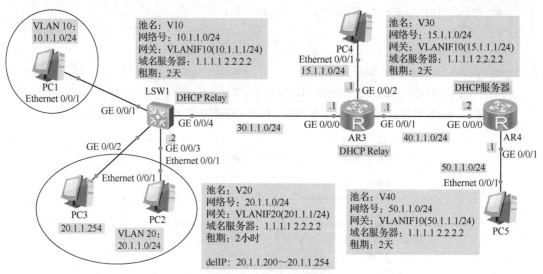

图 6-37　DHCP 拓展配置拓扑示意